FUNDAMENTALS OF SURVEYING

SAMPLE QUESTIONS & SOLUTIONS

FUNDAMENTALS OF SURVEYING
SAMPLE QUESTIONS & SOLUTIONS

Published by the

National Council of Examiners for Engineering and Surveying®

280 Seneca Creek Road, Clemson, SC 29631 800-250-3196 www.ncees.org

ISBN-13: 978-1-932613-20-9
ISBN-10: 1-932613-20-X

Printed in the United States of America

TABLE OF CONTENTS

UPDATES TO EXAMINATION INFORMATION

For current exam specifications, errata for this book, a list of calculators that may be used at the exam, guidelines for requesting special testing accommodations, and other information about exams, visit the NCEES Web site at www.ncees.org.

Introduction

One of the functions of the National Council of Examiners for Engineering and Surveying (NCEES) is to develop examinations that are taken by candidates for licensure as professional surveyors. NCEES has prepared this handbook to assist candidates who are preparing for the Fundamentals of Surveying (FS) examination. The NCEES is an organization established to assist and support the licensing boards that exist in all states and U.S. territories. NCEES provides these boards with uniform examinations that are valid measures of minimum competency related to the practice of surveying.

To develop reliable and valid examinations, NCEES employs procedures using the guidelines established in the *Standards for Educational and Psychological Testing* published by the American Psychological Association. These procedures are intended to maximize the fairness and quality of the examinations. To ensure that the procedures are followed, NCEES uses experienced testing specialists possessing the necessary expertise to guide the development of examinations using current testing techniques.

The examinations are prepared by committees composed of professional surveyors and surveying educators from throughout the nation. These surveyors supply the content expertise that is essential in developing examinations. By using the expertise of practitioners and educators, NCEES prepares examinations that are valid measures of minimum competency.

This book contains sample questions and solutions and is intended to be representative of the questions contained on an actual examination. The questions in this book are in accordance with the examination specifications although the number of questions is 50% of the number of questions on an actual examination. No representation is made or intended as to future examination questions, content, or subject matter.

Licensing Requirements

Eligibility
The primary purpose of licensure is to protect the public by evaluating the qualifications of candidates seeking licensure. While examinations offer one means of measuring the competency levels of candidates, most licensing boards also screen candidates based on education and experience requirements. Because these requirements vary between boards, it would be wise to contact the appropriate board. Board addresses and telephone numbers may be obtained by visiting our Web site at www.ncees.org or calling 800-250-3196.

Application Procedures and Deadlines
Application procedures for the examination and instructional information are available from individual boards. Requirements and fees vary among the boards, and applicants are responsible for contacting their board office. Sufficient time must be allotted to complete the application process and assemble required data.

DESCRIPTION OF EXAMINATIONS

Examination Schedule

The NCEES FS examination is offered to the boards in the spring and fall of each year. Dates of future administrations are published on the NCEES Web site at www.ncees.org. You should contact your board for specific locations of exam sites.

Examination Content

The 8-hour FS examination is a no-choice examination consisting of 170 multiple-choice questions. The examination is administered in two parts: 85 questions are administered in a 4-hour morning session, and 85 questions in a 4-hour afternoon session. Each question has four answer options. The examination specifications presented in this book give details of the subjects covered on the examination.

Typically, all information required to answer a question is provided within the statement of the question itself. However, there may be instances where two to three questions are grouped within a single scenario with common information appearing before the questions. In these instances, the questions supply any further information specific to the question and define what is expected as a response to the question. In many cases, the correct response requires a calculation and/or conclusion that demonstrates competent surveying judgment.

EXAMINATION DEVELOPMENT

Examination Validity

Testing standards require that the questions on a licensing examination be representative of the important knowledges and activities needed for competent practice in the profession. NCEES establishes the relationship between the examination questions and knowledges and activities by conducting an analysis of the profession that identifies the duties performed by the surveyor. This information is used to develop an examination content outline that guides the development of job-related questions.

The content of the current examinations is based on the analysis completed in 2003. The final report is titled *Report on the Test Specification Meeting for the NCEES Land Surveying Examinations*, by C. David Vale, Ph.D.

Examination Specifications
The examination content outline presented in this book specifies the subject areas that were identified for surveying and the percentage of questions devoted to each of them. The percentage of questions assigned to each of the subject areas reflects both the frequency and importance experienced in the practice of surveying.

Examination Preparation and Review
NCEES conducts examination development and review workshops twice annually to ensure that the questions remain current with respect to surveying practice. The content and format of the questions are reviewed by the committee members for compliance with the specifications and to ensure the quality and fairness of the examination. These surveyors are selected with the objective that they be representative of the profession in terms of geography, ethnic background, gender, and area of practice.

Minimum Competency
One of the most critical considerations in developing and administering examinations is the establishment of passing scores that reflect a standard of minimum competency. The concept of minimum competency is uppermost in the minds of the committee members as they assemble questions for the examination. Minimum competency, as measured by the examination component of the licensing process, is defined as the lowest level of knowledge at which a person can practice professional surveying in such a manner that will safeguard life, health, and property and promote the public welfare.

To accomplish the setting of fair passing scores that reflect the standard of minimum competency, NCEES conducts passing score studies on a periodic basis. At these studies, a representative panel of surveyors familiar with the candidate population uses a criterion-referenced procedure to set the passing score for the examination. Such procedures are widely recognized and accepted for occupational licensing purposes. The panel discusses the concept of minimum competency and develops a written standard of minimum competency that clearly articulates what skills and knowledge are required of a surveyor intern. Following this, the panelists take the examination and then evaluate the difficulty level of each question in the context of the standard of minimum competency.

NCEES does not use a fixed-percentage pass rate such as 70% or 75% because licensure is designed to ensure that practitioners possess enough knowledge to perform professional activities in a manner that protects the public welfare. The key issue is whether an individual candidate is competent to practice and not whether the candidate is better or worse than other candidates.

The passing score can vary from one administration of the examination to another to reflect differences in difficulty levels of the examinations. However, the passing score is always based on the standard of minimum competency. To avoid confusion that might arise from fluctuations in the passing score, scores are converted to a standard scale that adopts 70 as the passing score. This technique of converting to a standard scale is commonly employed by testing specialists.

SCORING PROCEDURES

The examination consists of 170 equally weighted multiple-choice questions. There is no penalty for marking incorrect responses; therefore candidates should answer each question on the examination. Only one response should be marked for each question. No credit is given where two or more responses are marked. The examination is compensatory—poor scores in some subjects can be offset by superior performance elsewhere.

EXAMINATION PROCEDURES AND INSTRUCTIONS

Visit the NCEES Web site for current information about exam procedures and instructions.

Examination Materials

Before the morning and afternoon sessions, proctors will distribute examination booklets containing an answer sheet. You should not open the examination booklet until you are instructed to do so by the proctor. Read the instructions and information given on the front and back covers, shown in Appendix A. Enter your name in the upper right corner of the front cover. Listen carefully to all the instructions the proctor reads.

The answer sheets for the multiple-choice questions are machine scored. For proper scoring, the answer spaces should be blackened completely. Since April 2002, NCEES has provided mechanical pencils with 0.7-mm HB lead to be used in the examination. You are not permitted to use any other writing instrument. If you decide to change an answer, you must erase the first answer completely. Incomplete erasures and stray marks may be read as intended answers. One side of the answer sheet is used to collect identification and biographical data. Proctors will guide you through the process of completing this portion of the answer sheet prior to taking the test. This process will take approximately 15 minutes.

Starting and Completing the Examination

You are not to open the examination booklet until instructed to do so by your proctor. If you complete the examination with more than 15 minutes remaining, you are free to leave after returning all examination materials to the proctor. Within 15 minutes of the end of the examination, you are required to remain until the end to avoid disruption to those still working and to permit orderly collection of all examination materials. Regardless of when you complete the examination, you are responsible for returning the numbered examination booklet assigned to you. Cooperate with the proctors collecting the examination materials. Nobody will be allowed to leave until the proctors have verified that all materials have been collected.

References

The FS examination is closed book. No references will be permitted. NCEES supplies, as part of the examination, a collection of reference formulas similar to that starting on page 17.

Calculators

Beginning with the April 2004 exam administration, the NCEES has strictly prohibited certain calculators from exam sites. Devices having a QWERTY keypad arrangement similar to a typewriter or keyboard are not permitted. Devices not permitted include but are not limited to palmtop, laptop, handheld, and desktop computers, calculators, databanks, data collectors, and organizers. The NCEES Web site (www.ncees.org) gives specific details on calculators.

Special Accommodations

The NCEES document *Guidelines for Requesting Religious and ADA Accommodations* explains the requirements for taking an NCEES exam with special testing accommodations. Candidates who wish to request special testing accommodations should refer to the NCEES Web site (www.ncees.org) under the "Exams" heading to find this document, along with frequently asked questions and forms for making the requests. To allow adequate evaluation time, NCEES must receive requests no later than 60 days prior to the exam administration.

EXAMINATION SPECIFICATIONS

FUNDAMENTALS OF SURVEYING (FS)
EXAMINATION SPECIFICATIONS

Effective October 2005

	Knowledge	Approximate Percentage of the Examination
I.	Algebra and Trigonometry	11%
II.	Higher Math (beyond trigonometry)	4%
III.	Probability and Statistics, Measurement Analysis, and Data Adjustment	5%
IV.	Basic Sciences	4%
V.	Geodesy, Survey Astronomy, and Geodetic Survey Calculation	6%
VI.	Computer Operations and Programming	6%
VII.	Written Communication	6%
VIII.	Boundary Law, Cadastral Law and Administration	13%
IX.	Business Law, Management, Economics, Finance, and Survey Planning Process and Procedures	6%
X.	Field Data Acquisition and Reduction	10%
XI.	Photo/Image Data Acquisition and Reduction	4%
XII.	Graphical Communication, Mapping	6%
XIII.	Plane Survey Calculation	10%
XIV.	Geographic Information System (GIS) Concepts	4%
XV.	Land Development Principles	5%

The exam contains 170 questions.

The fifteen knowledge areas on the FS examination and typical surveying activities associated with them are as follows.

I. Algebra and Trigonometry

(units of measurement; formula development; formula manipulation; solving systems of equations; basic mensuration formulas for length, area, volume; quadratic equations; trigonometric functions; right triangle solutions; oblique triangle solutions; spherical triangle solutions; trigonometric identities)

- Perform astronomic measurements.
- Perform trigonometric leveling.
- Perform differential leveling.
- Compute survey data.
- Compute areas and volumes.
- Determine and prepare lot and street patterns for land division.
- Design horizontal and vertical alignment for roads within a subdivision.

II. Higher Math (beyond trigonometry)

(analytic geometry; linear algebra; equation of a line, circle, parabola, ellipse; differentiation of functions; integration of elementary functions; infinite series; mathematical modeling)

- Perform geodetic surveys using conventional methods.
- Perform geodetic and/or plane surveys using GPS methods.
- Perform astronomic measurements.
- Compute survey data.
- Analyze and adjust survey data.
- Design horizontal and vertical alignment for roads within a subdivision.

III. Probability and Statistics, Measurement Analysis, and Data Adjustment

(standard deviation; variance; standard deviation of unit weight; tests of significance; concept of probability and confidence intervals; error ellipses; data distributions and histograms; analysis of error sources; error propagation; control network analysis; blunder trapping and elimination; least squares adjustment; calculation of uncertainty of position; accuracy standards; analysis of historical measurements)

- Determine levels of precision and order of accuracy.
- Perform geodetic and/or plane surveys using GPS methods.
- Compute survey data.
- Analyze and adjust survey data.
- Determine levels of precision and order of accuracy.
- Perform geodetic and/or plane surveys using conventional surveys.
- Perform geodetic and/or plane surveys using GPS methods.
- Perform astronomic measurements.
- Perform record or as-built surveys.
- Perform ALTA/ACSM surveys.
- Perform hydrographic surveys.
- Perform trigonometric leveling.
- Perform differential leveling.
- Perform photogrammetric control surveys.
- Produce survey data using photogrammetric methods.

- Perform boundary surveys.
- Perform route and right-of-way surveys.
- Perform topographic surveys.
- Perform flood plain surveys.
- Perform construction surveys.
- Perform condominium surveys.
- Compute survey data.
- Analyze and adjust survey data.
- Reconcile survey and record data.
- Convert survey data to an appropriate datum.
- Prepare work sheets for analysis of surveys.
- Determine locations of boundary lines and encumbrances.
- Determine and prepare lot and street patterns for land division.
- Design horizontal and vertical alignment for roads within a subdivision.
- Develop and/or provide data for LIS/GIS.

IV. Basic Sciences

(light and wave propagation; basic electricity; optics; gravity; refraction; mechanics; forces; kinematics; temperature and heat; biology; dendrology; geology; plant science)

- Calibrate instruments.

V. Geodesy, Survey Astronomy, and Geodetic Survey Calculation

(reference ellipsoids; gravity fields; geoid; geodetic datums; direction and distance on the ellipsoid; conversion from geodetic heights to elevation; orbit determination and tracking; determination of azimuth using common celestial bodies; time systems; calculation of position on a recognized coordinate system such as latitude/longitude; state plane coordinate systems; UTM coordinate systems; coordinate transformations; scale factors; meridian convergence)

- Select appropriate vertical and/or horizontal datum and basis of bearings.
- Perform geodetic surveys using conventional methods.
- Perform geodetic and/or plane surveys using GPS methods.
- Perform astronomic measurements.
- Perform hydrographic surveys.
- Perform differential leveling.
- Analyze and adjust survey data.
- Convert survey data to an appropriate datum.
- Determine levels of precision and order of accuracy.
- Perform record or as-built surveys.
- Perform ALTA/ACSM surveys.
- Perform trigonometric leveling.
- Perform photogrammetric control surveys.
- Produce survey data using photogrammetric methods.
- Perform boundary surveys.
- Perform route and right-of-way surveys.
- Perform topographic surveys.
- Perform flood plain surveys.
- Compute survey data.
- Prepare work sheets for analysis of surveys.

- Determine locations of boundary lines and encumbrances.
- Develop and/or provide data for LIS/GIS.

VI. Computer Operations and Programming

(operating systems; graphical user interfaces (Windows); data flow; bits and bytes; internet, computer architecture; programming a computer in a compiled language; order of arithmetic operations; programming concepts such as decision statements, flow charts, looping, arrays; spreadsheet operations)

- Compute survey data.
- Analyze and adjust survey data.
- Convert survey data to an appropriate datum.
- Utilize computer-aided drafting systems.

VII. Written Communication

(written communication; grammar; sentence structure; punctuation; bibliographical referencing)

- Evaluate project elements to define scope of work.
- Prepare and negotiate proposals and/or contracts.
- Consult and coordinate with allied professionals and/or regulatory agencies.
- Consult with and advise clients and/or their agents.
- Facilitate regulatory review and approval of project documents and maps.
- Determine and secure entry rights.
- Gather parol evidence.
- Perform boundary surveys.
- Advise clients regarding boundary uncertainties.
- Review documents with clients and/or attorneys.
- Prepare sketches and/or preliminary plats.
- Prepare survey maps, plats, and reports.
- Prepare land descriptions.

VIII. Boundary Law, Cadastral Law, and Administration

(land descriptions; real property rights; concepts of land ownership; case law; statute law; conveyancing; official records; land record sources; legal instruments of title; U.S. Public Land Survey System; colonial/metes and bounds survey system; subdivision survey system; other cadastral systems; rules of evidence relative to land boundaries and court appearance; boundary control and legal principles; order of importance of conflicting title elements; possession principles; conflict resolution; riparian/littoral water boundaries; boundary evidence; simultaneous and sequential conveyance)

- Facilitate regulatory review and approval of project documents and maps.
- Determine and secure entry rights.
- Research and evaluate evidence from private record sources.
- Research and evaluate evidence from public record sources.
- Research and evaluate court records and case law.
- Gather and evaluate parol evidence.
- Perform boundary surveys.
- Perform condominium surveys.
- Reconcile survey and record data.
- Identify and evaluate field evidence for possession, boundary line discrepancies, and potential adverse possession claims.
- Identify riparian and/or littoral boundaries.
- Apply Public Land and other Survey System principles.
- Evaluate the priority of conflicting title elements.
- Determine locations of boundary lines and encumbrances.
- Advise clients regarding boundary uncertainties.
- Testify as an expert witness.
- Review documents with clients and/or attorneys.
- Determine subdivision development requirements and constraints.
- Determine and prepare lot and street patterns for land division.
- Perpetuate and/or establish monuments and their records.
- Document potential possession claims.
- Prepare and file record of survey.
- Identify pertinent physical features, landmarks, and existing monumentation.
- Perform route and right-of-way surveys.
- Prepare survey maps, plats, and reports.
- Prepare land descriptions.

IX. Business Law, Management, Economics, Finance, Survey Planning Processes and Procedures

(sole proprietorships, corporations, partnership structures; contract law; tax structure; employment law; liability; operation analysis and optimization; land economics; appraisal science; critical path analysis; human resource management principles; cost/benefit analysis of a project or operation; econometric modeling; time value of money; budgeting; techniques for planning and conducting surveys including boundary surveys, control surveys, hydrographic surveys, topographic surveys, route surveys, aerial surveys, construction surveys; issues related to professional liability, ethics, and courtesy)

- Evaluate project elements to define scope of work.
- Prepare and negotiate proposals and/or contracts.
- Consult and coordinate with allied professionals and/or regulatory agencies.

- Consult with and advise clients and/or their agents.
- Facilitate regulatory review and approval of project documents and maps.
- Determine and secure entry rights.
- Advise clients regarding boundary uncertainties.
- Testify as an expert witness.
- Review documents with clients and/or attorneys.
- Document potential possession claims.
- Prepare survey maps, plats, and reports.
- Develop and/or provide data for LIS/GIS.

X. Field Data Acquisition and Reduction

(field notes and electronic data collection; measurement of distances, angles and directions; modern instruments and their construction and use; tapes; levels; theodolites; total stations; EDMs; GPS; hydrographic data collection instruments; construction layout instruments and procedures for routes and structures; historical measurement methods)

- Determine levels of precision and order of accuracy.
- Recover horizontal/vertical control.
- Identify pertinent physical features, landmarks, and existing monumentation.
- Calibrate instruments.
- Perform geodetic and/or plane surveys using conventional methods.
- Perform geodetic and/or plane surveys using GPS methods.
- Perform astronomic measurements.
- Perform record or as-built surveys.
- Perform ALTA/ACSM surveys.
- Perform hydrographic surveys.
- Perform trigonometric leveling.
- Perform differential leveling.
- Perform photogrammetric control surveys.
- Perform field verifications of photogrammetric maps.
- Produce survey data using photogrammetric methods.
- Perform boundary surveys.
- Perform route and right-of-way surveys.
- Perform topographic surveys.
- Perform flood plain surveys.
- Perform construction surveys.
- Perform condominium surveys.
- Perpetuate and/or establish monuments and their records.

XI. Photo/Image Data Acquisition and Reduction

(cameras; image scanners; digitizers; stereo plotters; photo and stereomodel orientation; orthophoto production; georectification; image processing; raster/vector data conversions)

- Determine levels of precision and order of accuracy.
- Perform record or as-built surveys.
- Perform ALTA/ACSM surveys.
- Perform photogrammetric control surveys.
- Perform field verifications of photogrammetric maps.
- Produce survey data using photogrammetric methods.

- Utilize survey data produced from photogrammetric methods.
- Perform topographic surveys.
- Perform flood plain surveys.
- Prepare survey maps, plats, and reports.

XII. Graphical Communication, Mapping

(principles of effective graphical display of spatial information; preparation of sketches; scaled drawings; survey plats and maps; interpretation of features on three-dimensional drawings; principles of cartography and map projections; computer mapping; use of overlays)

- Perform record or as-built surveys.
- Perform ALTA/ACSM surveys.
- Produce survey data using photogrammetric methods.
- Utilize survey data produced from photogrammetric methods.
- Prepare work sheets for analysis of surveys.
- Utilize computer-aided drafting systems.
- Determine and prepare lot and street patterns for land division.
- Design horizontal and vertical alignment for roads within a subdivision.
- Prepare sketches and/or preliminary plats.
- Prepare and file record of survey.
- Prepare survey maps, plats, and reports.
- Develop and/or provide data for LIS/GIS.

XIII. Plane Survey Calculation

(computation and adjustment of traverses; COGO computation of boundary lines, route alignments, construction layout, and subdivision plats; calculation of route curves and volumes)

- Determine levels of precision and order of accuracy.
- Calibrate instruments.
- Perform geodetic and/or plane surveys using conventional methods.
- Perform geodetic and/or plane surveys using GPS methods.
- Perform astronomic measurements.
- Perform record or as-built surveys.
- Perform ALTA/ACSM surveys.
- Perform hydrographic surveys.
- Perform trigonometric leveling.
- Perform differential leveling.
- Perform photogrammetric control surveys.
- Produce survey data using photogrammetric methods.
- Perform boundary surveys.
- Perform route and right-of-way surveys.
- Perform topographic surveys.
- Perform flood plain surveys.
- Perform construction surveys.
- Perform condominium surveys.
- Compute survey data.
- Analyze and adjust survey data.
- Reconcile survey and record data.
- Convert survey data to an appropriate datum.

- Prepare work sheets for analysis of surveys.
- Determine locations of boundary lines and encumbrances.
- Determine and prepare lot and street patterns for land division.
- Design horizontal and vertical alignment for roads within a subdivision.
- Develop and/or provide data for LIS/GIS.

XIV. Geographic Information System (GIS) Concepts

(spatial data storage, retrieval, and analysis systems; relational database systems; spatial objects; attribute value measurement; data definitions; schemas; metadata concepts; coding standards; GIS analysis of networks; buffering; overlay; spatial data accuracy standards)

- Utilize computer-aided drafting systems.
- Perpetuate and/or establish monuments and their records.
- Prepare and file records of surveys.
- Prepare survey maps, plats, and reports.
- Develop and/or provide data for LIS/GIS.

XV. Land Development Principles

(land planning and practices; laws controlling land use; drainage systems; construction methods; geometric and physical aspects of site analysis; design of land subdivisions; street alignment calculations; application of subdivision standards)

- Prepare sketches and/or preliminary plats.
- Prepare survey maps, plats, and reports.
- Prepare land descriptions.
- Evaluate project elements to define scope of work.
- Prepare and negotiate proposals and/or contracts.
- Consult and coordinate with allied professionals and/or regulatory agencies.
- Consult with and advise clients and/or their agents.
- Facilitate regulatory review and approval of project documents and maps.
- Determine subdivision development requirements and constraints.
- Determine and prepare lot and street patterns for land division.
- Perpetuate and/or establish monuments and their records.
- Design horizontal and vertical alignment for roads within a subdivision.

REFERENCE FORMULAS SUPPLIED IN EXAMINATION BOOKLET

CONVERSIONS AND OTHER USEFUL RELATIONSHIPS

* 1 U.S. survey foot = $\frac{12}{39.37}$ m

* 1 international foot = 0.3048 m

* 1 in. = 25.4 mm (international)

1 mile = 1.60935 km

* 1 acre = 43,560 ft^2 = 10 square chains

* 1 ha = 10,000 m^2 = 2.47104 acres

* 1 rad = $\frac{180^\circ}{\pi}$

1 kg = 2.2046 lb

1 L = 0.2624 gal

1 ft^3 = 7.481 gal

1 gal of water weighs 8.34 lb

1 ft^3 of water weighs 62.4 lb

1 atm = 29.92 in. Hg = 14.696 psi

Gravity acceleration (g) = 9.807 m/s^2 = 32.174 ft/sec^2

Speed of light in a vacuum (c) = 299,792,458 m/s = 186,282 miles/sec

°C = (°F – 32)/1.8

1 min of latitude (ϕ) ≅ 1 nautical mile

1 nautical mile = 6,076 ft

Mean radius of the earth ≅ 20,906,000 ft ≅ 6,372,000 m

* Denotes exact value. All others correct to figures shown.

METRIC PREFIXES		
Multiple	Prefix	Symbol
10^{-18}	atto	a
10^{-15}	femto	f
10^{-12}	pico	p
10^{-9}	nano	n
$\mathbf{10^{-6}}$	**micro**	**μ**
10^{-3}	milli	m
10^{-2}	centi	c
10^{-1}	deci	d

METRIC PREFIXES		
Multiple	Prefix	Symbol
10^{1}	deka	da
$\mathbf{10^{2}}$	**hecto**	**h**
10^{3}	kilo	k
10^{6}	mega	M
10^{9}	giga	G
10^{12}	tera	T
$\mathbf{10^{15}}$	**peta**	**P**
10^{18}	exa	E

QUADRATIC EQUATION

$ax^2 + bx + c = 0$

$$\text{Roots} = \frac{-b \pm \sqrt{b^2 - 4ac}}{2a}$$

OBLIQUE TRIANGLES

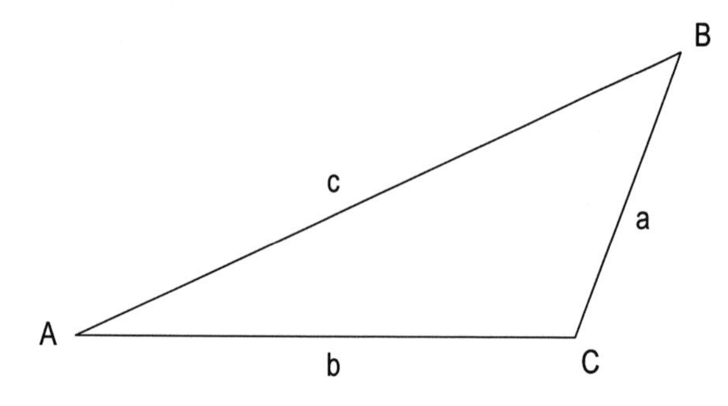

Law of sines

$$\frac{a}{\sin A} = \frac{b}{\sin B} = \frac{c}{\sin C}$$

Law of cosines

$$a^2 = b^2 + c^2 - 2bc\cos A$$

or

$$\cos A = \frac{b^2 + c^2 - a^2}{2bc}$$

$$\text{Area} = \frac{ab\sin C}{2}$$

$$\text{Area} = \frac{a^2 \sin B \sin C}{2\sin A}$$

$$\text{Area} = \sqrt{s(s-a)(s-b)(s-c)}$$

where $s = (a + b + c)/2$

SPHERICAL TRIANGLES

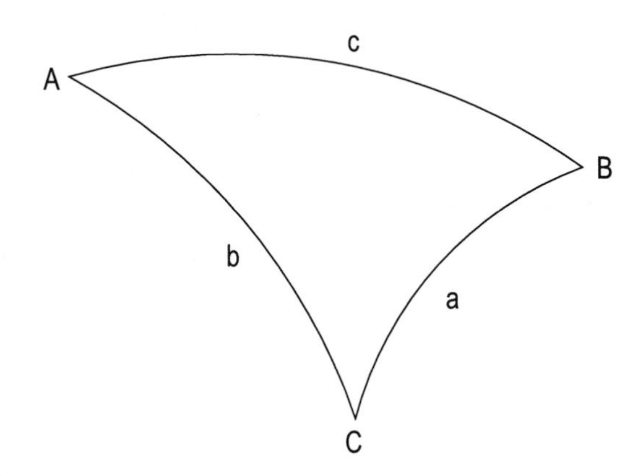

Law of sines

$$\frac{\sin a}{\sin A} = \frac{\sin b}{\sin B} = \frac{\sin c}{\sin C}$$

Law of cosines

$$\cos a = \cos b \cos c + \sin b \sin c \cos A$$

$$\text{Area of sphere} = 4\pi R^2$$

$$\text{Volume of sphere} = \frac{4}{3}\pi R^3$$

$$\text{Spherical excess in sec.} = \frac{bc \sin A}{9.7 \times 10^{-6} R^2}$$

where R = mean radius of the earth

PROBABILITY AND STATISTICS

$$\sigma = \sqrt{\frac{\sum (x_i - \bar{x})^2}{n-1}} = \sqrt{\sum \frac{v^2}{n-1}}$$

where:

- σ = standard deviation (sometimes referred to as standard error)
- Σv^2 = sum of the squares of the residuals (deviation from the mean)
- n = number of observations
- $\bar{x}$ = mean of the observations (individual measurements x_i)

$$\sigma_{sum} = \sqrt{\sigma_1^2 + \sigma_2^2 + \ldots + \sigma_n^2}$$

$$\sigma_{series} = \sigma\sqrt{n}$$

$$\sigma_{mean} = \frac{\sigma}{\sqrt{n}}$$

$$\sigma_{product} = \sqrt{A^2\sigma_b^2 + B^2\sigma_a^2}$$

$$\Sigma = \begin{bmatrix} \sigma_x^2 & \sigma_{xy} \\ \sigma_{xy} & \sigma_y^2 \end{bmatrix}$$

$$\tan 2\theta = \frac{2\sigma_{xy}}{\sigma_x^2 - \sigma_y^2}$$ where θ = the counter clockwise angle from the x axis

Relative weights are inversely proportional to variances, or:

$$W_a \propto \frac{1}{\sigma_a^2}$$

Weighted mean:

$$\overline{M}_w = \frac{\sum WM}{\sum W}$$

where:

- M_w = weighted mean
- ΣWM = sum of individual weights times their measurements
- ΣW = sum of the weights

HORIZONTAL CIRCULAR CURVES

D = Degree of curve, arc definition
L = Length of curve from P.C. to P.T.
c = Length of sub-chord
ℓ = Length of arc for sub-chord
d = Central angle for sub-chord

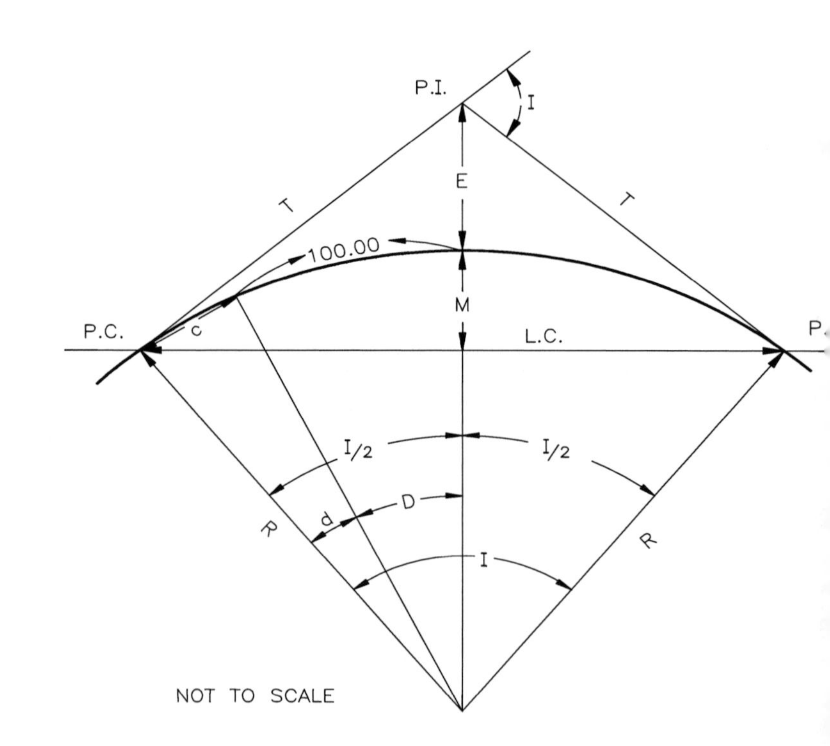

$$D = \frac{5,729.58}{R}$$

$$T = R \tan(I/2)$$

$$L = RI\frac{\pi}{180} = \frac{I}{D}(100)$$

$$LC = 2R \ \sin(I/2)$$

$$c = 2R \sin(d/2)$$

$$d = \ell D/100$$

$$M = R\left[1 - \cos(I/2)\right]$$

$$E = R\left[\frac{1}{\cos(I/2)} - 1\right]$$

$$\text{Area of sector} = \frac{RL}{2} = \frac{\pi R^2 I}{360}$$

$$\text{Area of segment} = \frac{\pi R^2 I}{360} - \frac{R^2 \sin I}{2}$$

$$\text{Area between curve and tangents} = R(T - L/2)$$

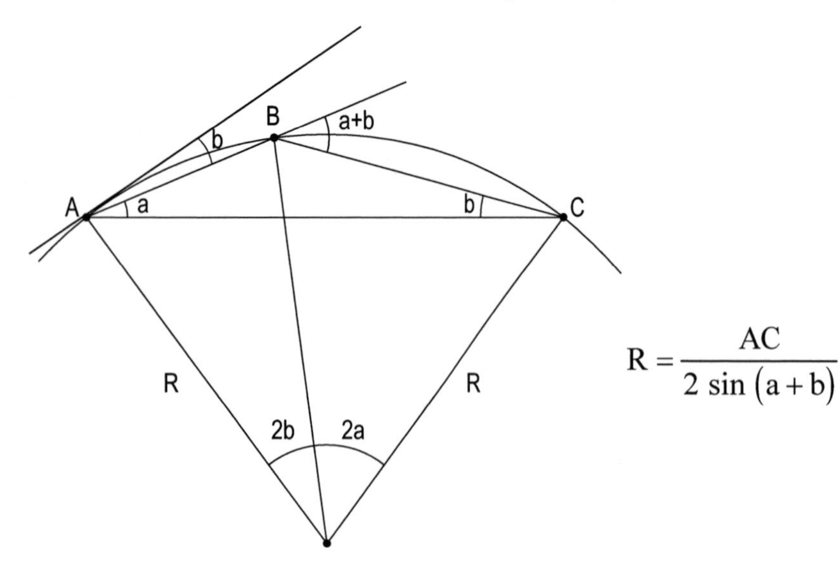

$$R = \frac{AC}{2 \sin(a+b)}$$

VERTICAL CURVE FORMULAS

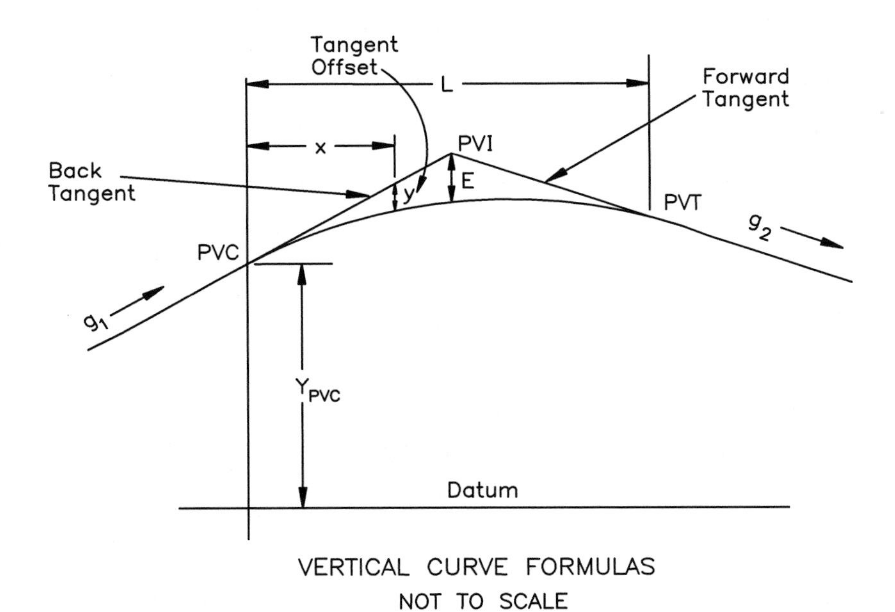

VERTICAL CURVE FORMULAS
NOT TO SCALE

L = Length of curve (horizontal)

PVC = Point of vertical curvature

PVI = Point of vertical intersection

PVT = Point of vertical tangency

g_1 = Grade of back tangent

g_2 = Grade of forward tangent

x = Horizontal distance from PVC (or point of tangency) to point on curve

a = Parabola constant

y = Tangent offset

E = Tangent offset at PVI

r = Rate of change of grade

Tangent elevation = $Y_{PVC} + g_1 x$
and = $Y_{PVI} + g_2 (x - L/2)$

Curve elevation = $Y_{PVC} + g_1 x + ax^2$
= $Y_{PVC} + g_1 x + [(g_2 - g_1)/(2L)]x^2$

$$y = ax^2; \qquad a = \frac{g_2 - g_1}{2L};$$

$$E = a\left(\frac{L}{2}\right)^2; \qquad r = \frac{g_2 - g_1}{L}$$

Horizontal distance to min/max elevation on curve, $x_m = -\dfrac{g_1}{2a} = \dfrac{g_1 L}{g_1 - g_2}$

ASTRONOMY

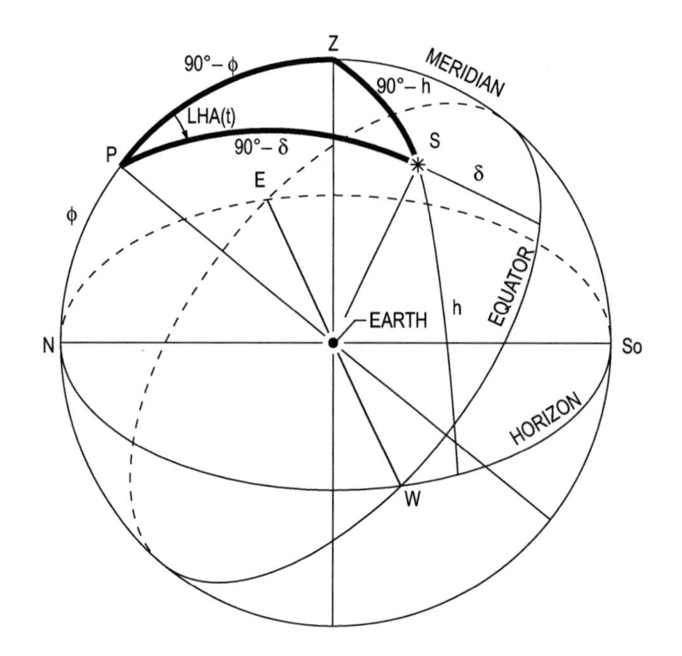

Cos (Az) = (sin δ – sin ϕ sin h)/(cos ϕ cos h) (altitude method)

Tan (Az) = –sin (LHA)/(cos ϕ tan δ – sin ϕ cos LHA) (hour angle method)

Sin h = sin ϕ sin δ + cos ϕ cos δ cos LHA

t = LHA or 360° – LHA

Horizontal circle correction for sun's semi-diameter = SD/cos h

Equations accurate for Polaris only:

$$h = \phi + p \cos LHA$$

$$Az = -(p \sin LHA)/\cos h$$

where:

Az	=	Azimuth (from north) to sun/star
δ	=	Declination
ϕ	=	Latitude
h	=	Altitude of sun/star
LHA	=	Local hour angle (sometimes referred to as "t" or "hour angle")
SD	=	Arc length of sun's semi-diameter
p	=	Polar distance of Polaris

PHOTOGRAMMETRY

$$\text{Scale} = \frac{ab}{AB} = \frac{f}{H-h}$$

(vertical photograph)

$$\text{Relief displacement} = \frac{rh}{H}$$

(vertical photograph)

Parallax equations:

$$p = x - x'$$

$$X = \frac{xB}{p}$$

$$Y = \frac{yB}{p}$$

$$h = H - \frac{fB}{p}$$

$$h_2 = h_1 + \frac{(p_2 - p_1)}{p_2}(H - h_1)$$

where:

f	=	Focal length
h	=	Height above datum
H	=	Flying height above datum
r	=	Radial distance from principal point
p	=	Parallax measured on stereo pair
B	=	Airbase of stereo pair
x, y	=	Coordinates measured on left photo
x′	=	Coordinate measured on right photo
X, Y	=	Ground coordinates

Lens equation:

$$\frac{1}{o} + \frac{1}{i} = \frac{1}{f}$$

where:

o	=	Object distance
i	=	Image distance
f	=	Focal length

Snell laws:

$$n \sin \phi = n' \sin \phi'$$

where:

n	=	Refractive index
ϕ	=	Angle of incidence

Curvature and refraction:

$$(c + r) = 0.0206M^2$$

where:

(c + r)	=	Combined effect of curvature and refraction in feet
M	=	Distance in thousands of feet

GEODESY

Ellipsoid

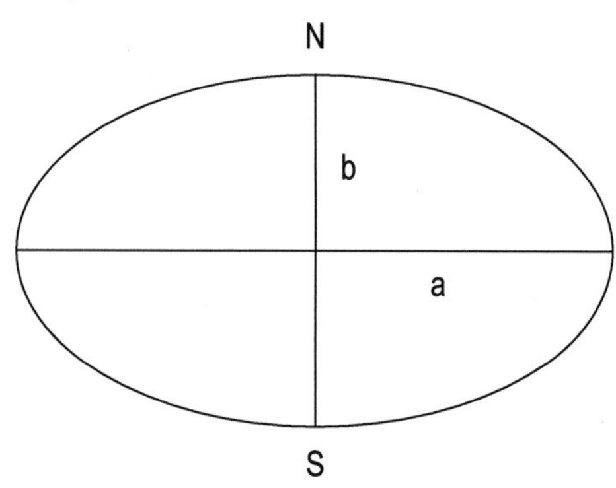

a = semi-major axis

b = semi-minor axis

$$\text{Flattening, } f = \frac{a - b}{a}$$

$$(\text{usually published as } 1/f)$$

$$\text{Eccentricity, } e^2 = \frac{a^2 - b^2}{a^2}$$

$$\text{Radius in meridian, } M = \frac{a(1 - e^2)}{(1 - e^2 \sin\phi)^{3/2}}$$

$$\text{Radius in prime vertical, } N = \frac{a}{(1 - e^2 \sin^2\phi)^{1/2}}$$

Angular convergence of meridians

$$\theta_{rad} = \frac{d \tan\phi (1 - e^2 \sin^2\phi)^{1/2}}{a}$$

Linear convergence of meridians

$$= \frac{\ell d \tan (1 - e^2 \sin^2\phi)^{1/2}}{a}$$

where:

ϕ = Latitude

d = Distance along parallel at latitude ϕ

ℓ = Length along meridians separated by d

Ellipsoid definitions:

GRS80: a = 6,378,137.0 m
1/f = 298.25722101

Clark 1866: a = 6,378,206.4 m
1/f = 294.97869821

Orthometric correction:

Correction = $-0.005288 \sin 2\phi h \Delta\phi \operatorname{arc} 1'$

where: ϕ = latitude at starting point

h = datum elevation in meters or feet at starting point

$\Delta\phi$ = change in latitude in minutes between the two points (+ in the direction of increasing latitude or towards the pole)

STATE PLANE COORDINATES

Scale factor = Grid distance/geodetic (ellipsoidal) distance

Elevation factor = R/(R + H +N)

where:

R = Ellipsoid radius

H = Orthometric height

N = Geoid height

For precision less than 1/200,000:

R = 20,906,000 ft

H = Elevation above sea level

N = 0

ELECTRONIC DISTANCE MEASUREMENT

$V = c/n$

$\lambda = V/f$

$$D = \frac{(m\lambda + d)}{2}$$

where:

V = Velocity of light through the atmosphere (m/s)

c = Velocity of light in a vacuum

n = Index of refraction

λ = Wave length (m)

f = Modulated frequency in hertz (cycles/sec)

D = Distance measured

m = Integer number of full wavelengths

d = Fractional part of the wavelength

ATMOSPHERIC CORRECTION

A 10°C temperature change or a pressure difference of 1 in. of mercury produces a distance correction of approximately 10 parts per million (ppm).

AREA FORMULAS

Area by coordinates where i is point order in a closed polygon.

$$\text{Area} = \frac{1}{2}\left[\sum_{i=1}^{n} X_i Y_{i+1} - \sum_{i=1}^{n} X_i Y_{i-1}\right]$$

Trapezoidal Rule

$$\text{Area} = w\left(\frac{h_1 + h_n}{2} + h_2 + h_3 + h_4 + \ldots + h_{n-1}\right)$$

Simpson's 1/3 Rule

$$\text{Area} = w\left[h_1 + 2\left(\Sigma h_{odds}\right) + 4\left(\Sigma h_{evens}\right) + h_n\right]/3$$

EARTHWORK FORMULAS

Average end area formula

$\text{volume} = L(A_1 + A_2)/2$

Prismoidal formula

$\text{volume} = L(A_1 + 4A_m + A_2)/6$

Pyramid or cone

$\text{volume} = h(\text{Area of Base})/3$

TAPE CORRECTION FORMULAS

Correction for temperature

$C_t = 6.5 \times 10^{-6}\,(T - T_s)L$

Correction for tension

$C_p = (P - P_s)L/(AE)$

Correction for sag

$C_s = (w^2 l^3)/(24P^2)$

where:

T = Temperature of tape during measurement, °F

T_s = Temperature of tape during calibration, °F

L = Distance measured, ft

P = Pull applied during measurement, lb

P_s = Pull applied during calibration, lb

A = Cross-sectional area of tape, in^2

E = Modulus of elasticity of tape, psi

w = Weight of tape, lb/ft

l = Length of unsupported span, ft

STADIA

Horizontal distance = $KS \cos^2 \alpha$

Vertical distance = $KS \sin \alpha \cos \alpha$

where:

K = Stadia interval factor (usually 100)

S = Rod intercept

α = Slope angle measured from horizontal

UNIT NORMAL DISTRIBUTION TABLE

x	f(x)	F(x)	R(x)	2R(x)	W(x)
0.0	0.3989	0.5000	0.5000	1.0000	0.0000
0.1	0.3970	0.5398	0.4602	0.9203	0.0797
0.2	0.3910	0.5793	0.4207	0.8415	0.1585
0.3	0.3814	0.6179	0.3821	0.7642	0.2358
0.4	0.3683	0.6554	0.3446	0.6892	0.3108
0.5	0.3521	0.6915	0.3085	0.6171	0.3829
0.6	0.3332	0.7257	0.2743	0.5485	0.4515
0.7	0.3123	0.7580	0.2420	0.4839	0.5161
0.8	0.2897	0.7881	0.2119	0.4237	0.5763
0.9	0.2661	0.8159	0.1841	0.3681	0.6319
1.0	0.2420	0.8413	0.1587	0.3173	0.6827
1.1	0.2179	0.8643	0.1357	0.2713	0.7287
1.2	0.1942	0.8849	0.1151	0.2301	0.7699
1.3	0.1714	0.9032	0.0968	0.1936	0.8064
1.4	0.1497	0.9192	0.0808	0.1615	0.8385
1.5	0.1295	0.9332	0.0668	0.1336	0.8664
1.6	0.1109	0.9452	0.0548	0.1096	0.8904
1.7	0.0940	0.9554	0.0446	0.0891	0.9109
1.8	0.0790	0.9641	0.0359	0.0719	0.9281
1.9	0.0656	0.9713	0.0287	0.0574	0.9426
2.0	0.0540	0.9772	0.0228	0.0455	0.9545
2.1	0.0440	0.9821	0.0179	0.0357	0.9643
2.2	0.0355	0.9861	0.0139	0.0278	0.9722
2.3	0.0283	0.9893	0.0107	0.0214	0.9786
2.4	0.0224	0.9918	0.0082	0.0164	0.9836
2.5	0.0175	0.9938	0.0062	0.0124	0.9876
2.6	0.0136	0.9953	0.0047	0.0093	0.9907
2.7	0.0104	0.9965	0.0035	0.0069	0.9931
2.8	0.0079	0.9974	0.0026	0.0051	0.9949
2.9	0.0060	0.9981	0.0019	0.0037	0.9963
3.0	0.0044	0.9987	0.0013	0.0027	0.9973
Fractiles					
1.2816	0.1755	0.9000	0.1000	0.2000	0.8000
1.6449	0.1031	0.9500	0.0500	0.1000	0.9000
1.9600	0.0584	0.9750	0.0250	0.0500	0.9500
2.0537	0.0484	0.9800	0.0200	0.0400	0.9600
2.3263	0.0267	0.9900	0.0100	0.0200	0.9800
2.5758	0.0145	0.9950	0.0050	0.0100	0.9900

t-DISTRIBUTION TABLE

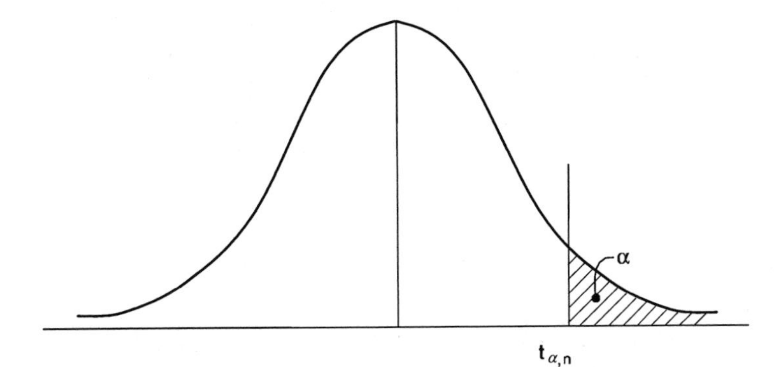

VALUES OF $t_{\alpha,n}$

n	α = 0.10	α = 0.05	α = 0.025	α = 0.01	α = 0.005	n
1	3.078	6.314	12.706	31.821	63.657	1
2	1.886	2.920	4.303	6.965	9.925	2
3	1.638	2.353	3.182	4.541	5.841	3
4	1.533	2.132	2.776	3.747	4.604	4
5	1.476	2.015	2.571	3.365	4.032	5
6	1.440	1.943	2.447	3.143	3.707	6
7	1.415	1.895	2.365	2.998	3.499	7
8	1.397	1.860	2.306	2.896	3.355	8
9	1.383	1.833	2.262	2.821	3.250	9
10	1.372	1.812	2.228	2.764	3.169	10
11	1.363	1.796	2.201	2.718	3.106	11
12	1.356	1.782	2.179	2.681	3.055	12
13	1.350	1.771	2.160	2.650	3.012	13
14	1.345	1.761	2.145	2.624	2.977	14
15	1.341	1.753	2.131	2.602	2.947	15
16	1.337	1.746	2.120	2.583	2.921	16
17	1.333	1.740	2.110	2.567	2.898	17
18	1.330	1.734	2.101	2.552	2.878	18
19	1.328	1.729	2.093	2.539	2.861	19
20	1.325	1.725	2.086	2.528	2.845	20
21	1.323	1.721	2.080	2.518	2.831	21
22	1.321	1.717	2.074	2.508	2.819	22
23	1.319	1.714	2.069	2.500	2.807	23
24	1.318	1.711	2.064	2.492	2.797	24
25	1.316	1.708	2.060	2.485	2.787	25
26	1.315	1.706	2.056	2.479	2.779	26
27	1.314	1.703	2.052	2.473	2.771	27
28	1.313	1.701	2.048	2.467	2.763	28
29	1.311	1.699	2.045	2.462	2.756	29
∞	1.282	1.645	1.960	2.326	2.576	∞

CRITICAL VALUES OF THE F DISTRIBUTION – TABLE

For a particular combination of numerator and denominator degrees of freedom, entry represents the critical values of F corresponding to a specified upper tail area (α).

$\alpha = 0.05$ — 0 — $F(\alpha, df_1, df_2)$ — F

Denominator df_2	Numerator df_1 1	2	3	4	5	6	7	8	9	10	12	15	20	24	30	40	60	120	∞
1	161.4	199.5	215.7	224.6	230.2	234.0	236.8	238.9	240.5	241.9	243.9	245.9	248.0	249.1	250.1	251.1	252.2	253.3	254.3
2	18.51	19.00	19.16	19.25	19.30	19.33	19.35	19.37	19.38	19.40	19.41	19.43	19.45	19.45	19.46	19.47	19.48	19.49	19.50
3	10.13	9.55	9.28	9.12	9.01	8.94	8.89	8.85	8.81	8.79	8.74	8.70	8.66	8.64	8.62	8.59	8.57	8.55	8.53
4	7.71	6.94	6.59	6.39	6.26	6.16	6.09	6.04	6.00	5.96	5.91	5.86	5.80	5.77	5.75	5.72	5.69	5.66	5.63
5	**6.61**	**5.79**	**5.41**	**5.19**	**5.05**	**4.95**	**4.88**	**4.82**	**4.77**	**4.74**	**4.68**	**4.62**	**4.56**	**4.53**	**4.50**	**4.46**	**4.43**	**4.40**	**4.36**
6	5.99	5.14	4.76	4.53	4.39	4.28	4.21	4.15	4.10	4.06	4.00	3.94	3.87	3.84	3.81	3.77	3.74	3.70	3.67
7	5.59	4.74	4.35	4.12	3.97	3.87	3.79	3.73	3.68	3.64	3.57	3.51	3.44	3.41	3.38	3.34	3.30	3.27	3.23
8	5.32	4.46	4.07	3.84	3.69	3.58	3.50	3.44	3.39	3.35	3.28	3.22	3.15	3.12	3.08	3.04	3.01	2.97	2.93
9	5.12	4.26	3.86	3.63	3.48	3.37	3.29	3.23	3.18	3.14	3.07	3.01	2.94	2.90	2.86	2.83	2.79	2.75	2.71
10	**4.96**	**4.10**	**3.71**	**3.48**	**3.33**	**3.22**	**3.14**	**3.07**	**3.02**	**2.98**	**2.91**	**2.85**	**2.77**	**2.74**	**2.70**	**2.66**	**2.62**	**2.58**	**2.54**
11	4.84	3.98	3.59	3.36	3.20	3.09	3.01	2.95	2.90	2.85	2.79	2.72	2.65	2.61	2.57	2.53	2.49	2.45	2.40
12	4.75	3.89	3.49	3.26	3.11	3.00	2.91	2.85	2.80	2.75	2.69	2.62	2.54	2.51	2.47	2.43	2.38	2.34	2.30
13	4.67	3.81	3.41	3.18	3.03	2.92	2.83	2.77	2.71	2.67	2.60	2.53	2.46	2.42	2.38	2.34	2.30	2.25	2.21
14	4.60	3.74	3.34	3.11	2.96	2.85	2.76	2.70	2.65	2.60	2.53	2.46	2.39	2.35	2.31	2.27	2.22	2.18	2.13
15	**4.54**	**3.68**	**3.29**	**3.06**	**2.90**	**2.79**	**2.71**	**2.64**	**2.59**	**2.54**	**2.48**	**2.40**	**2.33**	**2.29**	**2.25**	**2.20**	**2.16**	**2.11**	**2.07**
16	4.49	3.63	3.24	3.01	2.85	2.74	2.66	2.59	2.54	2.49	2.42	2.35	2.28	2.24	2.19	2.15	2.11	2.06	2.01
17	4.45	3.59	3.20	2.96	2.81	2.70	2.61	2.55	2.49	2.45	2.38	2.31	2.23	2.19	2.15	2.10	2.06	2.01	1.96
18	4.41	3.55	3.16	2.93	2.77	2.66	2.58	2.51	2.46	2.41	2.34	2.27	2.19	2.15	2.11	2.06	2.02	1.97	1.92
19	4.38	3.52	3.13	2.90	2.74	2.63	2.54	2.48	2.42	2.38	2.31	2.23	2.16	2.11	2.07	2.03	1.98	1.93	1.88
20	**4.35**	**3.49**	**3.10**	**2.87**	**2.71**	**2.60**	**2.51**	**2.45**	**2.39**	**2.35**	**2.28**	**2.20**	**2.12**	**2.08**	**2.04**	**1.99**	**1.95**	**1.90**	**1.84**
21	4.32	3.47	3.07	2.84	2.68	2.57	2.49	2.42	2.37	2.32	2.25	2.18	2.10	2.05	2.01	1.96	1.92	1.87	1.81
22	4.30	3.44	3.05	2.82	2.66	2.55	2.46	2.40	2.34	2.30	2.23	2.15	2.07	2.03	1.98	1.94	1.89	1.84	1.78
23	4.28	3.42	3.03	2.80	2.64	2.53	2.44	2.37	2.32	2.27	2.20	2.13	2.05	2.01	1.96	1.91	1.86	1.81	1.76
24	4.26	3.40	3.01	2.78	2.62	2.51	2.42	2.36	2.30	2.25	2.18	2.11	2.03	1.98	1.94	1.89	1.84	1.79	1.73
25	**4.24**	**3.39**	**2.99**	**2.76**	**2.60**	**2.49**	**2.40**	**2.34**	**2.28**	**2.24**	**2.16**	**2.09**	**2.01**	**1.96**	**1.92**	**1.87**	**1.82**	**1.77**	**1.71**
26	4.23	3.37	2.98	2.74	2.59	2.47	2.39	2.32	2.27	2.22	2.15	2.07	1.99	1.95	1.90	1.85	1.80	1.75	1.69
27	4.21	3.35	2.96	2.73	2.57	2.46	2.37	2.31	2.25	2.20	2.13	2.06	1.97	1.93	1.88	1.84	1.79	1.73	1.67
28	4.20	3.34	2.95	2.71	2.56	2.45	2.36	2.29	2.24	2.19	2.12	2.04	1.96	1.91	1.87	1.82	1.77	1.71	1.65
29	4.18	3.33	2.93	2.70	2.55	2.43	2.35	2.28	2.22	2.18	2.10	2.03	1.94	1.90	1.85	1.81	1.75	1.70	1.64
30	**4.17**	**3.32**	**2.92**	**2.69**	**2.53**	**2.42**	**2.33**	**2.27**	**2.21**	**2.16**	**2.09**	**2.01**	**1.93**	**1.89**	**1.84**	**1.79**	**1.74**	**1.68**	**1.62**
40	4.08	3.23	2.84	2.61	2.45	2.34	2.25	2.18	2.12	2.08	2.00	1.92	1.84	1.79	1.74	1.69	1.64	1.58	1.51
60	4.00	3.15	2.76	2.53	2.37	2.25	2.17	2.10	2.04	1.99	1.92	1.84	1.75	1.70	1.65	1.59	1.53	1.47	1.39
120	3.92	3.07	2.68	2.45	2.29	2.17	2.09	2.02	1.96	1.91	1.83	1.75	1.66	1.61	1.55	1.50	1.43	1.35	1.25
∞	3.84	3.00	2.60	2.37	2.21	2.10	2.01	1.94	1.88	1.83	1.75	1.67	1.57	1.52	1.46	1.39	1.32	1.22	1.00

ECONOMICS

Factor Name	Converts	Symbol	Formula
Single Payment Compound Amount	to F given P	$(F/P, i\%, n)$	$(1+i)^n$
Single Payment Present Worth	to P given F	$(P/F, i\%, n)$	$(1+i)^{-n}$
Uniform Series Sinking Fund	to A given F	$(A/F, i\%, n)$	$\frac{i}{(1+i)^n-1}$
Capital Recovery	to A given P	$(A/P, i\%, n)$	$\frac{i(1+i)^n}{(1+i)^n-1}$
Uniform Series Compound Amount	to F given A	$(F/A, i\%, n)$	$\frac{(1+i)^n-1}{i}$
Uniform Series Present Worth	to P given A	$(P/A, i\%, n)$	$\frac{(1+i)^n-1}{i(1+i)^n}$
Uniform Gradient Present Worth	to P given G	$(P/G, i\%, n)$	$\frac{(1+i)^n-1}{i^2(1+i)^n}-\frac{n}{i(1+i)^n}$
Uniform Gradient † Future Worth	to F given G	$(F/G, i\%, n)$	$\frac{(1+i)^n-1}{i^2}-\frac{n}{i}$
Uniform Gradient Uniform Series	to A given G	$(A/G, i\%, n)$	$\frac{1}{i}-\frac{n}{(1+i)^n-1}$

Nomenclature and Definitions

A	Uniform amount per interest period
B	Benefit
BV	Book Value
C	Cost
d	Combined interest rate per interest period
D_j	Depreciation in year j
F	Future worth, value, or amount
f	General inflation rate per interest period
G	Uniform gradient amount per interest period
i	Interest rate per interest period
i_e	Annual effective interest rate
m	Number of compounding periods per year
n	Number of compounding periods; or the expected life of an asset
P	Present worth, value, or amount
r	Nominal annual interest rate
S_n	Expected salvage value in year n

Subscripts

j	at time j
n	at time n
†	$F/G = (F/A - n)/i = (F/A) \times (A/G)$

Nonannual Compounding

$$i_e = \left(1 + \frac{r}{m}\right)^m - 1$$

Book Value

BV = Initial cost – ΣD_j

Depreciation

Straight line $D_j = \frac{C - S_n}{n}$

Accelerated Cost Recovery System (ACRS)

D_j = (factor from table below) C

MODIFIED ACRS FACTORS				
Year	Recovery Period (Years)			
	3	5	7	10
	Recovery Rate (%)			
1	33.3	20.0	14.3	10.0
2	44.5	32.0	24.5	18.0
3	14.8	19.2	17.5	14.4
4	7.4	11.5	12.5	11.5
5		**11.5**	**8.9**	**9.2**
6		5.8	8.9	7.4
7			8.9	6.6
8			4.5	6.6
9				6.5
10				**6.5**
11				3.3

Capitalized Costs

Capitalized costs are present worth values using an assumed perpetual period of time.

Capitalized costs = $P = \frac{A}{i}$

SAMPLE QUESTIONS

FUNDAMENTALS OF SURVEYING SAMPLE QUESTIONS

1. One corner of a 60-ft × 120-ft lot, otherwise rectangular, is a curve with a radius of 20 ft and a central angle of 90°. The area (ft^2) of the lot is most nearly:

(A) 6,872
(B) 6,886
(C) 7,114
(D) 7,200

2. A small rectangular lot measures 120.00 ± 0.04 ft by 144.00 ± 0.05 ft. The area (ft^2) of the lot is best stated as:

(A) 17,280 ± 4.7
(B) 17,280 ± 8.3
(C) 17,280 ± 49.7
(D) 17,280 ± 87

GO ON TO THE NEXT PAGE

3. A client wishes to create a 1-acre parcel by establishing a North-South line, BC, as shown.

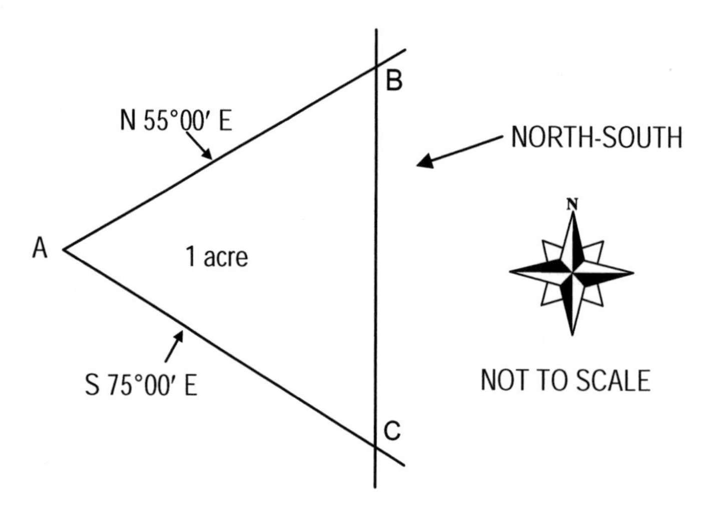

The length (ft) of Side AB is most nearly:

(A) 299.96
(B) 352.84
(C) 358.73
(D) 366.20

4. If $A = \begin{bmatrix} 3 & 4 & 7 \end{bmatrix}$ and $B = \begin{bmatrix} 5 \\ 0 \\ -2 \end{bmatrix}$ the matrix product AB is:

(A) $\begin{bmatrix} 15 \\ 0 \\ -14 \end{bmatrix}$

(B) $[1]$

(C) $\begin{bmatrix} 15 & 20 & 35 \\ 0 & 0 & 0 \\ -6 & -8 & -14 \end{bmatrix}$

(D) $\begin{bmatrix} 15 & 0 & -14 \end{bmatrix}$

5. The two survey lines on the figure are represented mathematically by two equations:

1. $\dfrac{E - E_1}{N - N_1} = \tan 45°$

2. $\dfrac{E - E_2}{N - N_2} = \tan 315°$

What is the correct equation for the east coordinate of the point of intersection?

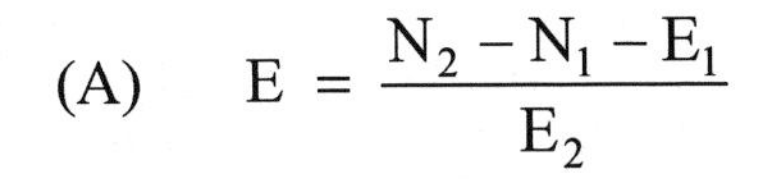

(A) $E = \dfrac{N_2 - N_1 - E_1}{E_2}$

(B) $E = \dfrac{N_1 - N_2}{2(E_1 + E_2)}$

(C) $E = \dfrac{N_1 + N_2}{E_1 + E_2}$

(D) $E = 1/2[(E_1 + E_2) - (N_1 - N_2)]$

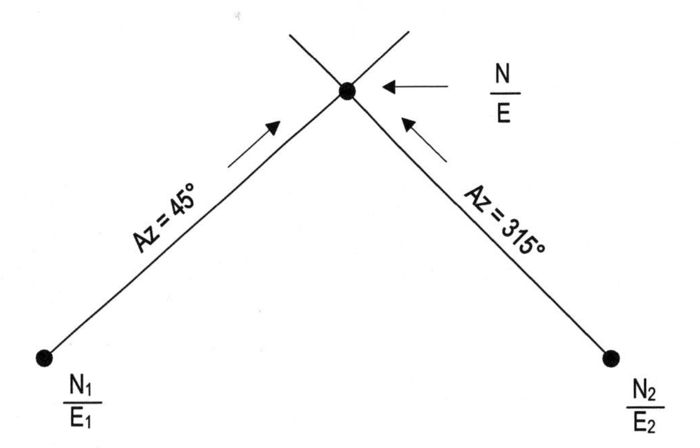

6. The length of a line is measured ten times by the same personnel with the same equipment under the same conditions. The lengths measured were:

215.86	215.78	215.84	215.82	215.86
215.80	215.84	215.83	215.86	215.81

The standard deviation of the set of measurements is most nearly:

(A) 0.023
(B) 0.025
(C) 0.027
(D) 0.032

GO ON TO THE NEXT PAGE

7. Two brass monuments set on a shady sidewalk have a known, verified horizontal separation of 99.96 ft. A surveyor measures between the monuments with a tape and reads 99.99 ft at a temperature of 83°F, holding a tension of 15 lb while the tape is fully supported.

The length (ft) of the surveyor's tape between the 0 and 100 marks while the temperature remains at 83°F is most nearly:

(A) 99.95
(B) 99.97
(C) 100.03
(D) The question cannot be answered with the information given.

8. In order to achieve the best possible horizontal accuracy, which of the following would be the most important with regard to the electronic distance measuring device used?

(A) Know the exact atmospheric temperature and pressure.

(B) Keep the battery fully charged.

(C) Use an umbrella to keep sun off the instrument.

(D) Avoid placing equipment beneath high-tension power lines.

9. You are planning a Polaris observation at latitude 42° North when the star is at or near upper culmination. You want to know the error in azimuth that will result from a 1-min timing error. You know that the meridian angle, t, is exactly 0° at upper culmination.

Use the approximations that the celestial sphere rotates once in approximately 24 hours, the declination of Polaris is approximately 89°10′, and the bearing of Polaris is approximated by the following equation:

$$Z = (\sin t)(p)/(\cos \text{latitude})$$

The azimuth error resulting from a 1-min timing error is most nearly:

(A) 04″
(B) 09″
(C) 15″
(D) 18″

GO ON TO THE NEXT PAGE

10. The figure of the earth considered as a sea level surface perpendicular at every point to the plumb line is known as the:

(A) spheroid
(B) Clarke ellipsoid of 1866
(C) geophysical surface
(D) geoid

11. For the program flow chart below, the final value of variable C is:

(A) 8
(B) 9
(C) 10
(D) 11

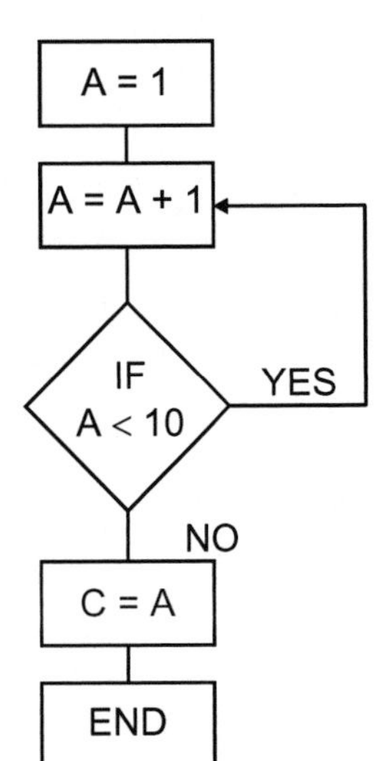

12. A typical 7 1/2-min USGS quadrangle (quad sheet) will show:

(A) state plane coordinate grid ticks at 10,000-ft intervals
(B) contours at an interval of 2 ft
(C) availability of aerial photographs
(D) all of the above

13. The legal term for information received from interviews with knowledgeable local residents that can be used to clarify or explain ambiguous words in a deed is:

(A) intrinsic evidence
(B) parol evidence
(C) colloquial evidence
(D) in sequitor localis

14. An assistant has prepared the following letter and asked you to review it.

Dear Mr. and Mrs. Brown:
Our company has been retaned to preform a boundry survey of land in you're area. When we are surveying this track of land, it maybe neccesary to enter your's to search for an locate monuments reletive to our survey. Please rest ensured that we will be carefull to do as little damage as possable.

We will also call you before coming, too.

Respectively,

R. Plumb, L.S.

How many grammatical and spelling mistakes are there?

(A) 8
(B) 12
(C) 15
(D) 18

15. The original government record of a fractional lot in the northwest quarter of Section 5 shows the following dimensions in chains: north side 19.83, east side 19.09, west side 19.31, and south side 20.14. The area (acres) of the lot on the original township plat would be most nearly:

(A) 38.33
(B) 38.35
(C) 38.37
(D) 38.39

16. The section corner G shown below is restored by double proportionate methods. Which of the following is correct?

(A) Line FG is parallel with Line AC, and Line EG is parallel with Line BD.

(B) Line FG is a cardinal direction, and Line EG is a cardinal direction.

(C) Line FG is perpendicular to Line BD, and Line GE is perpendicular to Line CA.

(D) The geometric relationship of lines FG and GE to lines BD and CA is not pertinent to the restoration of the section corner G.

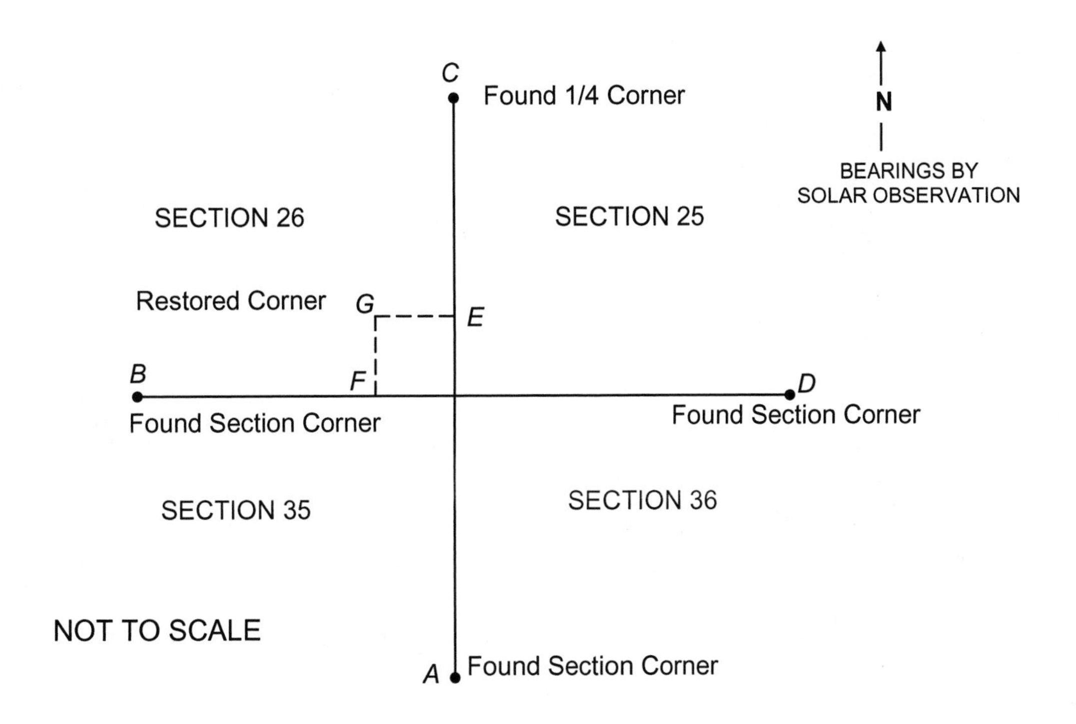

17. An easement is:

(A) always granted in writing

(B) a limited and non-possessory interest in a tract of land

(C) not an encumbrance on the lands of another

(D) always used for transportation or utility purposes

18. When you are researching records for a retracement survey for the right-of-way line of a road, railroad, or utility, the best source is the:

(A) National Geodetic Survey (NGS)

(B) original design and construction plans with the recorded legal descriptions

(C) actual stationing of the right-of-way as indicated by the construction and design plans

(D) deeds and surveys of adjacent property owners

19. A warranty deed is an example of:

(A) possession insurance

(B) a Torrens title

(C) a title guarantee

(D) an agreement between owners to fix a disputed boundary line

20. Where, from natural causes, land forms by imperceptible degrees upon the bank of a river or stream, the process and end result are called:

(A) accretion

(B) reliction

(C) revulsion

(D) erosion

21. While reviewing a record of survey performed by another surveyor in your area, you observe the comment "corner obliterated, recovered by collateral evidence." This statement means:

(A) all landowners affected by the corner location paid a fee to have it reestablished

(B) the corner was reestablished by evidence other than the monument or its accessories

(C) the corner was replaced by taking of oaths by the surveyor's assistants

(D) evidence to its location was gathered in a court of law

22. The description of Huffman's land ends with the phrase "together with an easement across Johnson's land for road purposes as shown . . . ." With respect to Huffman's description, the easement is:

(A) informative

(B) augmenting

(C) simultaneous

(D) encumbering

23. What is the "controlling call" in the following portion of a metes and bounds description?

". . . thence N 80°20′ E a distance of 800.28 ft to a 12-in. oak tree blazed on the south side, thence . . ."

(A) N 80°20′ E

(B) 800.28 ft

(C) the oak tree

(D) both (A) and (B) above

24. Which of the following statements about easements is **NOT** true?

(A) An easement can be created without the consent of the property owner.

(B) An easement can be valid if it is unrecorded.

(C) An easement cannot be conveyed separately and apart from the land it benefits.

(D) All existing easements on a parcel of land can be extinguished by mutual consent of the grantee and grantor during the conveyance of the parcel.

25. A surveyor should keep adequate files of survey records because:

(A) it is required by federal law

(B) otherwise a client can sue the surveyor

(C) it is a professional responsibility

(D) all of the above

26. Your survey company has decided to purchase a robotic total station field survey system for $40,000. The payment schedule recited in the loan agreement is $8,000 plus interest due payable at the end of each year. If the system is financed at 8% simple interest for 5 years, the additional cost to your company for financing versus paying cash when the system is delivered is most nearly:

(A) $2,000
(B) $3,200
(C) $9,600
(D) $16,000

27. The traditional world map, commonly used in grade schools, is based on a map projection originally devised by:

(A) Ellicot
(B) Lambert
(C) Washington
(D) Mercator

28. Which survey method was used to lay out a railroad centerline in the 1920s?

(A) Theodolite and tape
(B) Compass and tape
(C) Transit and chain
(D) Transit and tape

29. After leveling up a hill where backsight distances were taken at 200 ft and foresight distances at 150 ft, you discovered that the line of sight was inclined upward at 0.012 ft per 100-ft sight distance. The difference in elevation between the starting BM A and ending BM B was +50.035 ft. There were 20 instrument setups.

The adjusted elevation difference (ft) after correcting for line of sight inclination is most nearly:

(A) 49.915
(B) 50.023
(C) 50.029
(D) 50.155

GO ON TO THE NEXT PAGE

30. A survey party has set offset stakes for construction of the 8-in. sewer shown in the design plan below. When the existing 12-in. sewer line is uncovered for the construction of Maintenance Hole (MH) 1, it is found that the actual flow line elevation is 1,228.69 ft rather than the design elevation of 1,228.47 ft. The gradient must be revised, holding the flow line elevation of 1,229.27 ft at MH 2. If the elevation of the grade stake is 1,235.06 ft, the cut (ft) to the flow line that you would mark on the stake at Sta. 1+25 is most nearly:

(A) 5.98
(B) 6.08
(C) 6.18
(D) 6.25

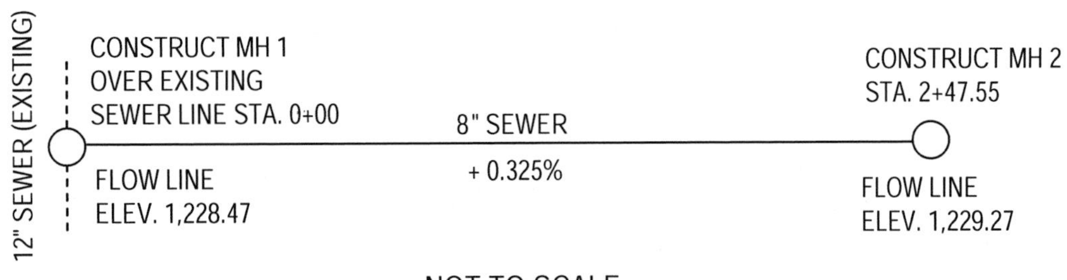

31. The distance on a vertical aerial photograph between two east-west hedge lines is measured and found to be 7.96 in. The hedge lines are approximately the north and south section lines of Section 16, which is regular. The terrain is approximately level. What is the approximate photo scale in the area between the two hedges?

(A) 1 : 663
(B) 1 : 24,000
(C) 1 in. = 663 ft
(D) 1 in. = 7,960 ft

GO ON TO THE NEXT PAGE

32. You plan to plot the following traverse on a sheet with dimensions of 18 in. wide × 24 in. long.

AB: S 0°25' E,	1,380.02 ft
BC: N 88°31' W,	2,495.00 ft
CD: N 0°25' W,	1,380.02 ft
DA: S 88 °31' E,	2,495.00 ft

The scale best suited to show maximum detail and to allow for a 1/2-in. margin is:

(A) 1:1,440
(B) 1:1,200
(C) 1:960
(D) 1:600

33. When is it proper to apply proportionate measurement to the location of property corners?

(A) To distribute a few feet of gap found to exist between a subdivision boundary and the boundary of the original tract

(B) When relocating lost corners in a platted subdivision

(C) When relocating lost corners in a sequence conveyance

(D) All of the above

34. The area of a lake is obtained by planimeter as 10 in^2 on a map at scale 1:50,000. The area (sq. mi.) covered by the lake is most nearly:

(A) 6.23
(B) 7.89
(C) 9.47
(D) 10.00

35. A traverse was run from Point A to Point E, and the coordinates of each point were computed with the following results:

Point	X Coordinate	Y Coordinate
A	100.00	100.00
B	250.55	232.66
C	388.26	95.98
D	466.15	2.15
E	609.50	−11.92

The distance and bearing, respectively, of a straight line from Point A to Point E are most nearly:

(A) 517.06 ft, S 09°54′ E

(B) 517.06 ft, S 80°06′ E

(C) 521.65 ft, S 12°23′ E

(D) 521.65 ft, S 77°37′ E

36. The coordinates of Point Q on a highway spiral relative to the TS are: X = 200, Y = 5. The deflection angle from the TS to Point Q is most nearly:

(A) 0°28′38″
(B) 1°25′55″
(C) 2°50′00″
(D) 4°05′02″

37. The following deflection angles were measured in a closed traverse:

P: 92°24′ R
Q: 150°42′ R
R: 15°37′ L
S: 132°35′ R

The balanced deflection angle at R is most nearly:

(A) 15°36′ L
(B) 15°37′ L
(C) 15°38′ L
(D) 15°39′ L

38. Direct and reverse zenith angles to a point are read as follows:

D = 36°12′18″
R = 323°47′36″

The vertical circle reading that must be set in the instrument to produce a vertical angle of 12°16′12″ is most nearly:

(A) 77°43′45″
(B) 77°43′48″
(C) 77°43′51″
(D) 77°43′54″

39. The state plane coordinate system differs from local plane systems in that:

(A) geodetic calculations must be made instead of plane calculations to determine positions
(B) astronomic north coincides with grid north in state plane systems
(C) distances are projected onto developable surfaces such as cones and cylinders rather than planes
(D) all grid meridian lines are parallel

40. An EDM distance of 1 mile is measured at an elevation of 1 mile. The earth's radius R is assumed to be 20,906,000 ft. The sea level distance (ft) is most nearly:

(A) 5,270.02
(B) 5,278.67
(C) 5,280.00
(D) 5,281.33

41. The elevation of BM A is 644.00 ft. A level in perfect adjustment is set midway between BM A and BM B. The backsight reading is 8.76 ft, and the foresight reading is 3.21 ft. If the level rod at BM B is held at an angle of 10° to the vertical, then the correct elevation (ft) of BM B is:

(A) 638.40
(B) 649.50
(C) 649.55
(D) 649.60

42. In using EDM equipment, better precision is attained when:

(A) long sights are taken

(B) the shortest possible sights are taken

(C) sights are taken as close to the ground as practical

(D) backsights and foresights are equal

GO ON TO THE NEXT PAGE

43. In attempting to restore the lost corner shown in the figure, the surveyor runs a connecting traverse between A, B, C, and D as follows:

Line	Balanced Length (ft)	True Bearing
AB	692.63	N 54°35′30″ W
BC	734.34	N 50°07′10″ E
CD	732.91	S 49°52′19″ E
DA	687.57	S 54°26′49″ W

The direction of AB is 305°24′30″ referenced to north.

If the coordinates of B are N 1,000.00, E 0.00, using double proportionate measurement, the coordinates of the lost corner are most nearly:

(A) N 1,000.00, E 554.40

(B) N 999.25, E 555.35

(C) N 1,030.49, E 555.34

(D) N 1,030.72, E 576.5

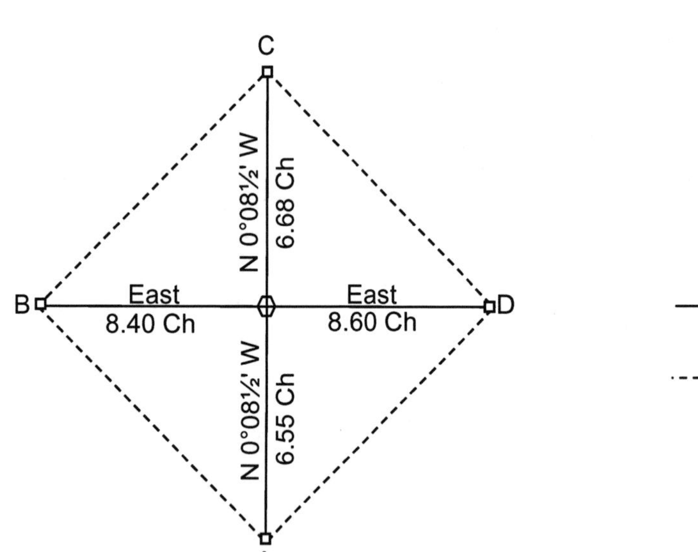

Lost Corner

Found Monument

Original Platted Lines

Contemporary Connecting Traverse Lines

GO ON TO THE NEXT PAGE

44. A level has been found to be out of adjustment. In making a two-peg test and adjustment, two stakes (A and B) were set 215 ft apart. When the instrument was set up midway between A and B, the rod readings were 4.26 on A and 5.15 on B. When the instrument was set up very close to A, the rod readings were 6.02 on A and 6.79 on B. With the instrument remaining near A, the crosshairs should be adjusted to a rod reading on B of most nearly:

(A) 6.67
(B) 6.79
(C) 6.85
(D) 6.91

45. Section 18 of T21N, R6W, was subdivided for the first time about 20 years ago. You wish to retrace that survey. The official distance shown in government notes for the north line of Section 18 is 78.39 chains. The measurement (chains) that should have been used for the north line of the NW 1/4 of the NW 1/4 (also called Lot 1) is most nearly:

(A) 18.39
(B) 19.20
(C) 19.60
(D) 38.39

46. A fundamental concept underlying the design of most vector-based GIS is:

(A) separate storage of spatial and attribute data components

(B) the use of relational databases for all data storage

(C) the integration of CAD and network databases

(D) the use of four number scales for attribute data

47. A clear zone navigation easement with a 34:1 slope begins at ground level 200 ft from the end of an active airport runway. The natural ground slope moving away from a point 200 ft from the end of the runway is 0.5% in an uphill direction. At what distance (ft) from the end of the runway can a 35-ft-tall structure be located and not violate the clear zone easement?

(A) 1,434
(B) 1,593
(C) 1,634
(D) 1,675

48. A topographic survey of a rectangular 1/2-acre city park to a scale of 1:120 could best be completed by:

(A) a grid system at 10-ft intervals

(B) a radial method using a total station

(C) low-altitude photogrammetry

(D) rapid static GPS

49. The specifications for a survey require that the boundaries of a subdivision be referenced to the state plane coordinate system and directions of the boundary lines be referenced to grid north. You can obtain a list of control monuments and coordinates from the:

(A) American Congress on Surveying and Mapping

(B) National Bureau of Standards

(C) National Geodetic Survey

(D) Bureau of Management

50. You have obtained a **current** vertical value on an NGS benchmark. This value will be on which of the following datums?

(A) NAVD 1929

(B) NAVD 1988

(C) NGVD

(D) NAD 27

51. The sum of the exterior angles of an eight-sided figure is most nearly:

(A) 1,800°
(B) 1,440°
(C) 1,080°
(D) none of the above

52. The center of a circle with a radius of 4 is at x = 5, y = –2. The equation of the circle is:

(A) $(x-5)^2 + (y-2)^2 - 4 = 0$

(B) $(x+5)^2 + (y+2)^2 - 4 = 0$

(C) $(x-5)^2 + (y+2)^2 - 16 = 0$

(D) $(x-5)^2 + (y+2)^2 + 16 = 0$

53. An angle is measured with a 1″ theodolite twelve times with the following results:

223°14′56″
223°14′52″
223°14′58″
223°14′59″
223°14′53″
223°14′55″
223°15′02″
223°15′00″
223°14′58″
223°14′59″
223°14′55″
223°14′54″

The standard deviation of the **MEAN** is most nearly:

(A) ±0.9″
(B) ±1.5″
(C) ±2.8″
(D) ±3.3″

54. Most modern electronic distance meters (EDM) utilize a collimated beam of infrared or near-infrared light. Infrared light is a:

(A) longer wavelength than visible light
(B) shorter wavelength than visible light
(C) variety of laser
(D) specific type of microwave

55. Which of the following computer components or peripherals often communicates with the computer using a parallel interface?

(A) Modem
(B) Mouse
(C) Printer
(D) Monitor

56. Consider the following sentence.

Atlanta Georgia is where the survey was done.

How many commas should be inserted into the above sentence?

(A) None
(B) One
(C) Two
(D) Three

57. Which of the following is **NOT** a common type of business ownership?

(A) Sole proprietorship
(B) Limited proprietorship
(C) Corporation
(D) Partnership

58. In determining overall performance of a total station, an accurate log should be kept of its operation. A log also provides historical legal verification of the instrument. Which of the following should always be recorded?

1. Name of stations where observations are made
2. Instrument and reflector model and serial number
3. Date and time of observation
4. Atmospheric observations
5. Weather conditions

(A) 1, 2, 3, 4, 5

(B) 1, 2, 5 only

(C) 2, 3, 4 only

(D) 3, 4, 5 only

59. When establishing horizontal control targets for a stereo photogrammetric survey, targets should be:

(A) as large as possible to aid in finding them

(B) placed on subdivision boundary corners, road intersection points, and important land corners

(C) placed in a straight line approximating the center of the flight line

(D) spaced at intervals that will assure their appearance on adjoining photos

60. A planimetric map differs from topographic maps in that the planimetric map:

(A) has no positional data

(B) shows vertical positions of points

(C) shows horizontal and vertical positions

(D) shows only the horizontal positions of points

61. A highway grade of +2.18% has an elevation of 432.76 at Station 157+34.55.

The grade elevation at Station 134+42.90 is most nearly:

(A) 282.80
(B) 382.80
(C) 432.26
(D) 482.72

62. Which of the following items is **NOT** a part of the data quality report in the spatial data transfer standards (SDTS)?

(A) Completeness

(B) Lineage

(C) Positional accuracy

(D) Data dictionary

63. The purpose of a spiral curve is to:

(A) elevate the inside of the curve

(B) decrease the amount of earthwork

(C) make the centerline fit the topography better

(D) decrease sudden development of centrifugal force

64. Assuming the earth is a sphere and its radius is 5,631 miles, the latitude of a point that is 2,700 miles north of the equator is most nearly:

(A) 27°47′26″ N

(B) 27°28′22″ N

(C) 21°36′46″ N

(D) 21°22′03″ N

65. Which of the following is an equation of an ellipse?

(A) $x^2 - 4x + 2y - 4 = 0$
(B) $x^2 + 4x - 2y - 4 = 0$
(C) $2x^2 + 4x - 3y^2 + 12y - 16 = 0$
(D) $2x^2 - 4x + 3y^2 + 12y - 16 = 0$

66. A thermometer, which is known to read 3°F too high, records a temperature of 46°F. The correct temperature is most nearly:

(A) 6.1°C
(B) 7.8°C
(C) 9.4°C
(D) 25.2°C

67. In developing a surveying-related computer program, which variable type would be most appropriate for representing coordinate values?

(A) Integer
(B) Boolean
(C) Character
(D) Floating point

68. Consider the following sentence:

The deed describes the objects bounding the premices, but parole evidence is usually resorted to, for the purpose of identifying the objects themselves.

How many words are spelled **INCORRECTLY?**

(A) One
(B) Two
(C) Three
(D) Four

GO ON TO THE NEXT PAGE

69. You are to set slope stakes along the roads within a subdivision. At Station 2+00 the finish grade elevation is 110.31 at the edge of the road, and the distance from the centerline to the edge of the road is 12.0 ft. The rod is being held at a distance of 28.5 ft from the centerline and rod reading is 12.1 ft while the H.I. is 119.77 ft. Typical cut and fill sections are shown below.

Your next step would be to:

(A) move in about 6 ft and try again

(B) move in about 10 ft and try again

(C) move out about 6 ft and try again

(D) drive in a stake since you are at the slope stake

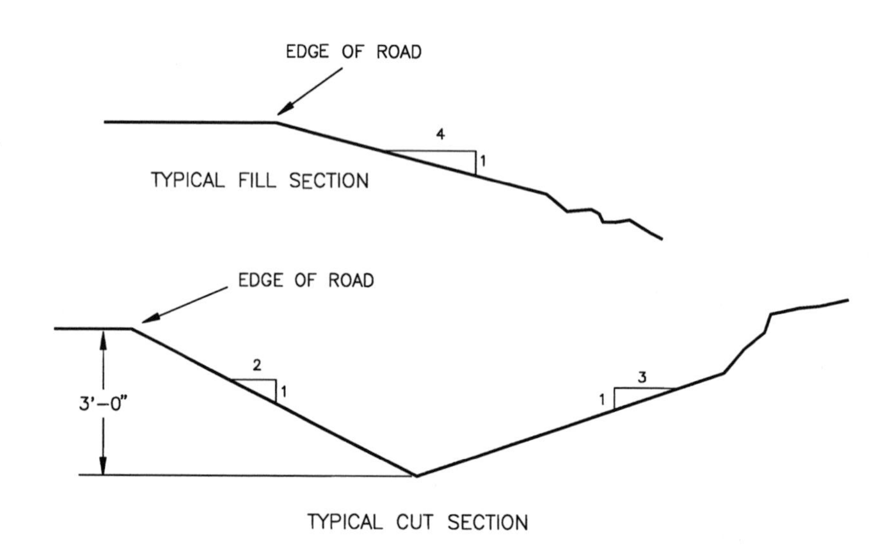

70. On an aerial photograph, the measured distance between two points is 5.134 in. On a 7.5-min topographic map (1:24,000 scale) the measured distance between these same two points is 1.689 in. The nominal scale ratio of the photo is most nearly:

(A) 1:658
(B) 1:7896
(C) 1:7920
(D) 1:24,000

71. In writing a description for a parcel of land lying to the northwest of a proposed highway right-of-way, you must compute the arc length along the right-of-way line from the PC to where this line crosses the property line. The length (ft) of this arc is most nearly:

(A) 618.04
(B) 621.32
(C) 637.92
(D) 641.81

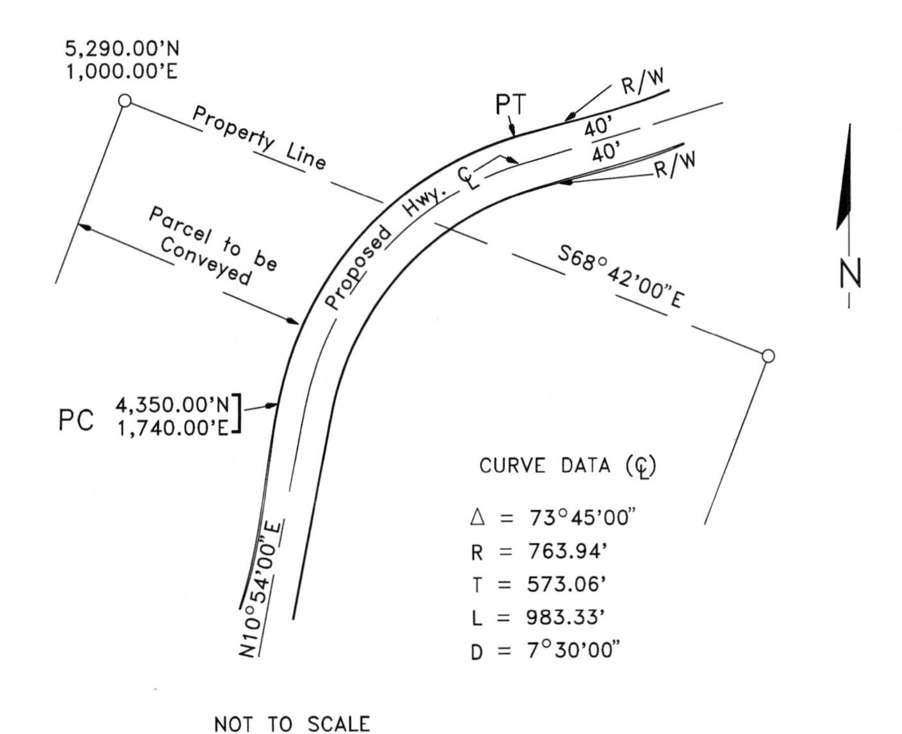

72. Which of the following data would likely be found in a county LIS project and would be valuable in preparing preliminary conceptual land development plans?

1. Flood plain boundaries
2. Utility information
3. Zoning
4. Location of boundary monuments
5. Overlaps and gaps
6. Soil data

(A) 1, 2, 3, 4 only

(B) 1, 2, 3, 6 only

(C) 1, 2, 4, 5 only

(D) 3, 4, 5, 6 only

73. A field crew has measured the spot elevations on a 50-ft grid that covers the 200-ft × 200-ft lot shown on the figure below. The client needs over 50% of the lot to be above an elevation of 650 ft in order to develop the lot. The percentage of the lot that is above the 650-ft elevation is most nearly:

(A) 35%
(B) 50%
(C) 65%
(D) 80%

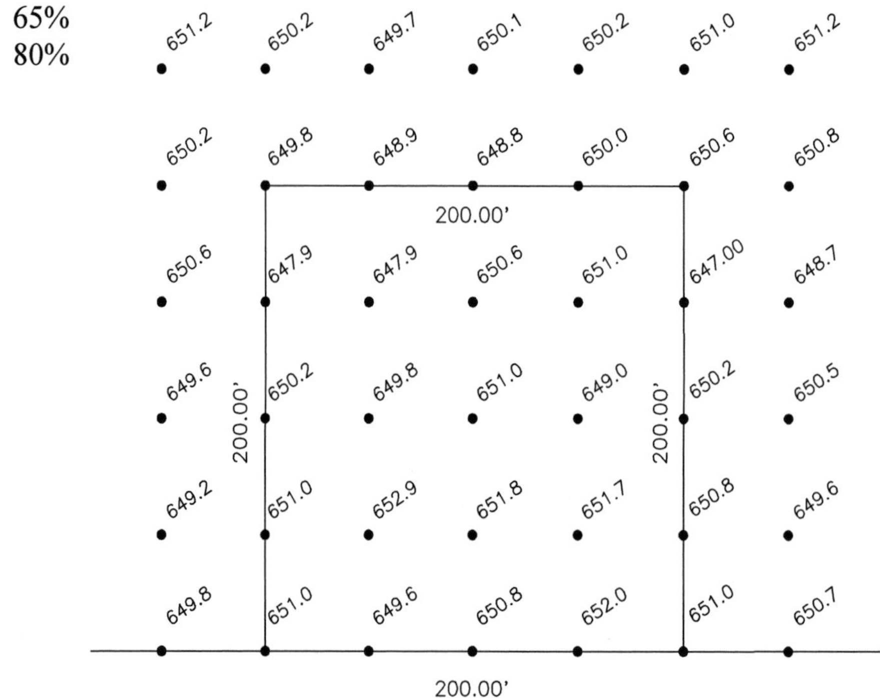

GO ON TO THE NEXT PAGE

74. A theodolite is used to sight on a 1-in.-diameter range pole at a distance of 172 ft. If the edge of the range pole is sighted instead of the center, the angular error introduced in the direction of the line is most nearly:

(A) 35″
(B) 50″
(C) 70″
(D) 100″

75. A thin-walled tank is constructed as a body of revolution of a parabola, as shown in the figure below. The base diameter is 20 ft, and the height of the tank is 25 ft. The volume (ft^3) of water in the tank when full is most nearly:

(A) $\frac{500}{3}\pi$

(B) $\frac{625}{2}\pi$

(C) $\frac{625}{2}\left(9 - 4\sqrt{2}\right)\pi$

(D) $\frac{6,875}{6}\pi$

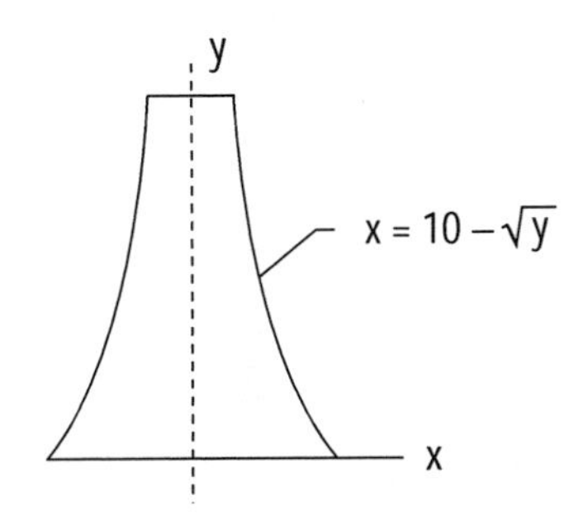

76. Light traveling through the atmosphere is "bent" by the change in air density. This refraction makes objects that are on the horizon appear:

(A) higher than actual

(B) lower than actual

(C) higher and wider than actual

(D) lower and wider than actual

77. A bit is the smallest item of data in a computer. How many bits make up a byte?

(A) 4
(B) 8
(C) 16
(D) 64

78. The intersection of the vertical plane containing the nadir point and the optical axis of the lens with the plane of the photograph is the:

(A) swing

(B) axis of tilt

(C) tilt

(D) principal line

79. An arc definition curve has D = 40°, I = 120°, P.C. = 9+60. The deflection angle from the P.C. to 10+00 is:

(A) 16°
(B) 11°
(C) 10°
(D) 8°

80. You have been asked to prepare a staking plan for a proposed bike trail. You are provided with the survey cross-sections at Stations 0+50 and 1+00 as shown below. Also provided is the typical design cross-section. Assume uniform slopes between cross-section elevations. Station 0+75 has a centerline design elevation of 785.0 ft.

The original ground slope between the hinge point and flow line at Station 0+75 is most nearly:

(A) –10.0%
(B) –5.0%
(C) –2.5%
(D) 0%

CENTERLINE

Station / CL Elev.	10'	10'	10'	10'	10'
1+00 / 792.1	792.4	790.1	790.1	791.5	792.7
0+50 / 791.2	792.4	789.9	788.9	790.3	791.4

EXISTING GROUND ELEVATIONS AT CROSS SECTIONS

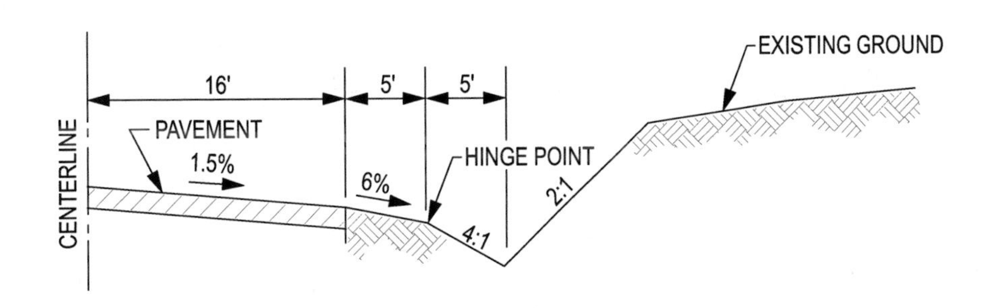

TYPICAL CROSS SECTION
NOT TO SCALE

81. A –4.0% grade and a +2.0% grade meet between Stations 60+00 and 70+00. An 800-ft-long, equal tangent, vertical curve will join the two grades.

Station	Elevation (ft)	Grade (%)
60+00	1,166.00	–4.0
70+00	1,159.00	+2.0

The vertical distance in feet (external of the curve) from the PVI to the midpoint of the curve is most nearly:

(A) 3.00
(B) 4.00
(C) 5.00
(D) 6.00

82. Which object described below will subtend the greatest angle at your eye?

(A) A tree 18 feet tall at 100 yards away
(B) A house 12 feet tall at 180 feet away
(C) A 1/2-inch diameter coin at 10 inches away
(D) The 2,170-mile diameter moon at 240,000 miles away

83. Compared to light at the blue end of the visible spectrum, light at the red end of the visible spectrum has:

(A) less heat
(B) a higher frequency
(C) a larger amplitude
(D) a longer wavelength

84. Consider the following equation:

A = B*C + D/C^2

where:

B = 2
C = 0.5
D = 127

The following notation applies to this question:

* = multiply
/ = divide
^ = raise to exponent

If the equation were executed by a spreadsheet or computer, the value of A would be most nearly:

(A) 509
(B) 512
(C) 130,050
(D) 299,081

GO ON TO THE NEXT PAGE

85. An island is formed by the intersections of Birch, Oak and Ash Streets. Specific details of the intersection are shown in the figure below. The length (ft) of the right-of-way line along the Birch Street side of the island is most nearly:

(A) 46.95
(B) 47.35
(C) 47.70
(D) 47.90

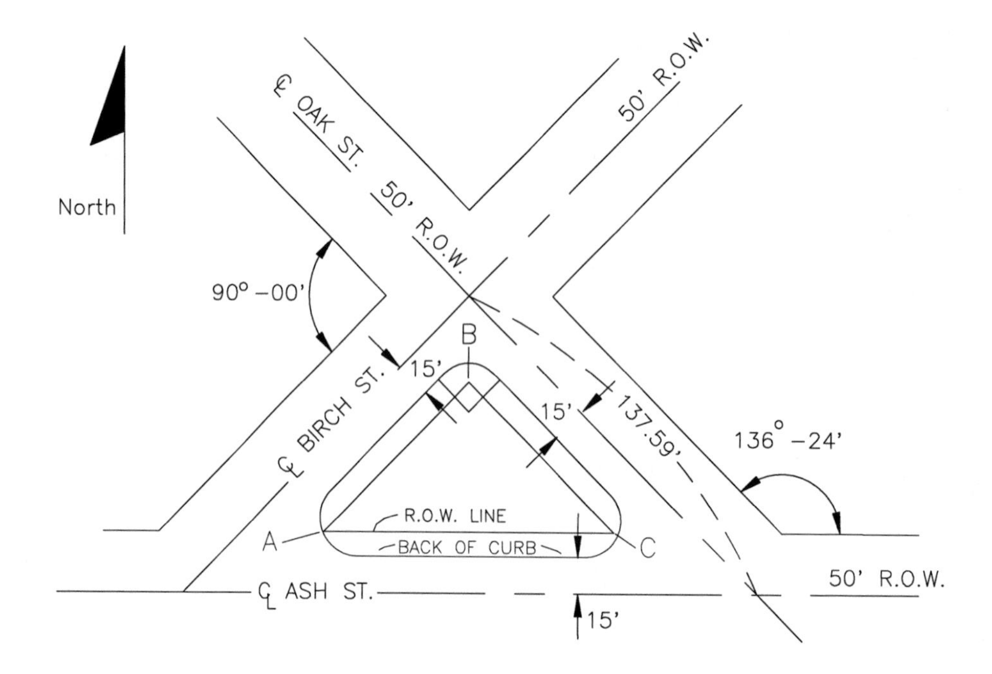

SOLUTIONS TO SAMPLE QUESTIONS

FUNDAMENTALS OF SURVEYING SOLUTIONS

1. (Reference: General mathematics reference)

Area = total area – area outside arc

$$\text{Area} = (60)(120) - \left[(20)(20) - \frac{(3.14)(20^2)}{4}\right]$$
$$= 7{,}200 - [400 - 314.2]$$
$$= 7{,}114.2 \text{ ft}^2$$

THE CORRECT ANSWER IS: (C)

2. (Reference: See List A, List D)

The equation for the error in a product AB, where σ_A and σ_B are the respective errors in A and B, is

$$\text{Error} = \pm\sqrt{(A\sigma_B)^2 + (B\sigma_A)^2}$$
$$= \pm\sqrt{(144 \times 0.04)^2 + (120 \times 0.05)^2}$$
$$= \pm\sqrt{33.18 + 36.00}$$
$$= \pm 8.32$$

THE CORRECT ANSWER IS: (B)

3. (Reference: See List A and general trigonometry reference and reference formulas supplied in the examination booklet.)

Parcel to contain 1.00 acre.

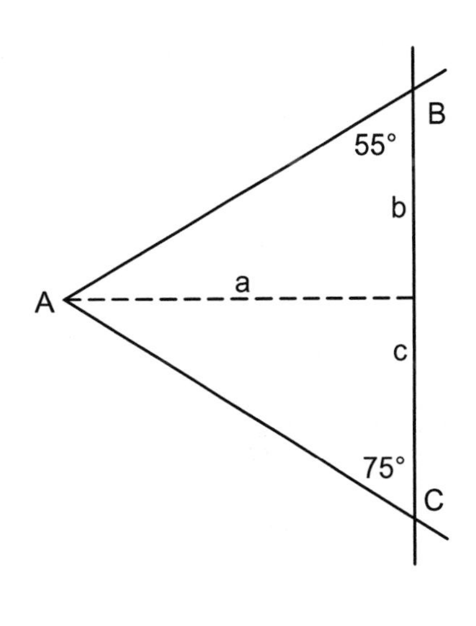

$$\frac{(b+c)(a)}{2} = 43{,}560$$
$$(b+c)(a) = 87{,}120$$
$$ab + ac = 87{,}120$$
$$(a)\left(\frac{a}{\tan 55^\circ}\right) + (a)\left(\frac{a}{\tan 75^\circ}\right) = 87{,}120$$
$$\frac{a^2}{\tan 55^\circ} + \frac{a^2}{\tan 75^\circ} = 87{,}120$$
$$\frac{5.16a^2}{5.329} = 87{,}120$$
$$a^2 = 89{,}973.35$$
$$a = 299.96$$
$$AB = 299.96 / \sin 55^\circ = 366.18 \text{ ft}$$

THE CORRECT ANSWER IS: D

4. (Reference: General mathematics reference)

$$AB = [3 \times 5 + 4 \times 0 + 7 \times (-2)] = [15 + 0 + (-14)] = [1]$$

(the result is a 1 × 1 matrix)

THE CORRECT ANSWER IS: (B)

5. (Reference: See List A or general mathematics reference)

tan 45° = 1
tan 315° = –1

Substituting these values into the two equations and rearranging gives the following:

$$E - E_1 = N - N_1 \quad (1)$$
$$E - E_2 = -N + N_2 \quad (2)$$

Add the two equations:

$$2E - (E_1 + E_2) = -(N_1 - N_2)$$

Rearrange:

$$E = 1/2\ [(E_1 + E_2) - (N_1 - N_2)]$$

THE CORRECT ANSWER IS: (D)

6. (Reference: See List A or List D)

$$\text{Mean} = \bar{x} = \frac{\Sigma x_i}{n} = \frac{2{,}158.3}{10} = 215.83\ \text{ft}$$

$$\text{Std. dev.} = \sqrt{\frac{\sum_i (x_i - \bar{x})^2}{n-1}}$$
$$= \sqrt{\frac{0.0068}{9}}$$
$$= \sqrt{0.000756}$$
$$= 0.027\ \text{ft}$$

THE CORRECT ANSWER IS: (C)

7. (Reference: See List A)

$$\frac{99.96}{99.99} = \frac{x}{100}$$

x = 99.97 ft

THE CORRECT ANSWER IS: (B)

8. (Reference: See List A)

THE CORRECT ANSWER IS: (A)

9. (Reference: See List A)

Z = (sin t) × (p)/cos ϕ p = 0°50′ ϕ = 42°
(For 1-min error in timing of transit/culmination)

t = 1 min of time, as it is the hour angle at the pole

$$1\text{ min} = \frac{1}{60}\text{ hour} = \frac{1}{60} \times \frac{1}{24} = \frac{1}{1,400}\text{ day} = \frac{360°}{1,440} = 0.25°$$

$$Z = \frac{(\sin 0.25°) \times (0°50')}{\cos 42°} = (0.005871) \times (0°50')$$

$$= 17.6''$$

THE CORRECT ANSWER IS: (D)

10. (Reference: See List A)

THE CORRECT ANSWER IS: (D)

11. (Reference: Basic computing reference)

The diamond-shaped box has two possible exits. The only way to get to the end and establish a final value of variable C is when the test of the statement IF A < 10 is not true. This first occurs when A = 10. This invokes the statement C = A and C has the value 10.

THE CORRECT ANSWER IS: (C)

12. (Reference: See List A or List C)

THE CORRECT ANSWER IS: (A)

13. (Reference: See List F, BLM Manual)

THE CORRECT ANSWER IS: (B)

14. (Reference: See general reference on grammar and a dictionary)

The grammatical and spelling mistakes are underlined.

Dear Mr. and Mrs. Brown:
Our company has been retaned to preform a boundry survey of land in you're area. When we are surveying this track of land, it maybe neccesary to enter your's to search for an locate monuments reletive to our survey. Please rest ensured that we will be carefull to do as little damage as possable.

We will also call you before coming, too.

Respectively,

R. Plumb, L.S.

Corrections for each mistake are as follows:

1. retained
2. perform
3. boundary
4. your
5. tract
6. may be (2 words)
7. necessary
8. yours
9. and
10. relative
11. assured
12. careful
13. possible
14. "also" and "too" (redundant)
15. Respectfully

THE CORRECT ANSWER IS: (C)

15. (Reference: See List F, BLM Manual)

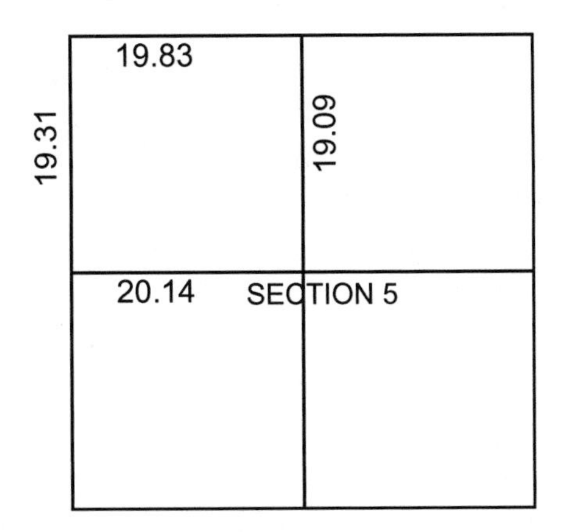

$$\left(\frac{19.31+19.09}{2}\right)\left(\frac{19.83+20.14}{2}\right)$$
$$= 383.71 \text{ sq ch.}$$

$$A = \frac{383.71}{10 \text{ sq ch / acre}}$$

$$= 38.37 \text{ acres}$$

THE CORRECT ANSWER IS: (C)

16. (Reference: See List F, BLM Manual)

THE CORRECT ANSWER IS: (B)

17. (Reference: See List H, *Black's Law Dictionary*)

THE CORRECT ANSWER IS: (B)

18. (Reference: See List F and List H)

THE CORRECT ANSWER IS: (B)

19. (Reference: See List H, definitions)

THE CORRECT ANSWER IS: (C)

20. (Reference: See List F)

THE CORRECT ANSWER IS: (A)

21. (Reference: See List H, *Black's Law Dictionary*)

THE CORRECT ANSWER IS: (B)

22. (Reference: See List H, definitions, and List F)

THE CORRECT ANSWER IS: (B)

23. (Reference: See List F)

THE CORRECT ANSWER IS: (C)

24. (Reference: See List H, *Black's Law Dictionary*)

THE CORRECT ANSWER IS: (A)

25. (Reference: See List F and List H)

THE CORRECT ANSWER IS: (C)

26. (Reference: See basic reference on economics)

Year	Amount Owed Each Year	Simple Interest @ 8%
1	40,000	3,200
2	32,000	2,560
3	24,000	1,920
4	16,000	1,280
5	8,000	640
	Total Interest Paid	$9,600

THE CORRECT ANSWER IS: (C)

27. (Reference: See List C)

THE CORRECT ANSWER IS: (D)

28. (Reference: See List A)

In the 1920s, transit and tape were dominant.

THE CORRECT ANSWER IS: (D)

29. (Reference: See List A)

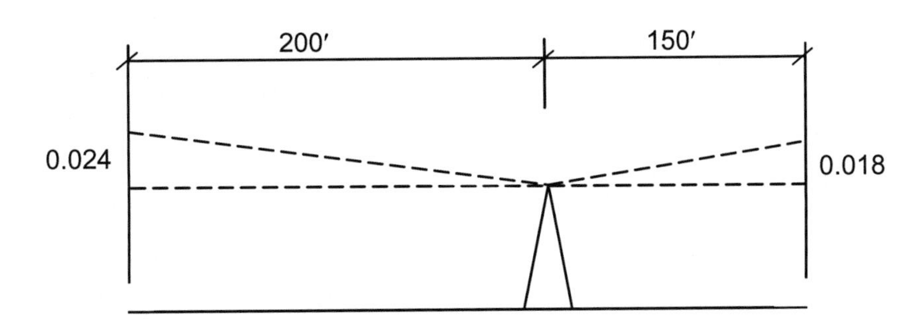

The line of sight is inclined 0.012 ft/100 ft.
$(0.012 \times 2) - (0.012 \times 1.5) = 0.006$

Each setup has an error of 0.006 ft.
With 20 setups, total = (20) (0.006) = 0.120

Ending elevation = 50.035 – 0.120 = 49.915

THE CORRECT ANSWER IS: (A)

30. (Reference: See List B)

Determine the revised gradient: (1,229.27 – 1,228.69)/247.55 = 0.00234

Determine the flow line elevation at Station 1+25: 1,228.69 + (125)(0.00234) = 1,228.98

Determine the cut at Station 1+25: 1,235.06 – 1,228.98 = 6.08 ft

THE CORRECT ANSWER IS: (B)

31. (Reference: See List A)

7.96 in. represents 5,280 ft
Therefore, 1 in. represents 5,280/7.96 = 663 ft

THE CORRECT ANSWER IS: (C)

32. (Reference: See List A)

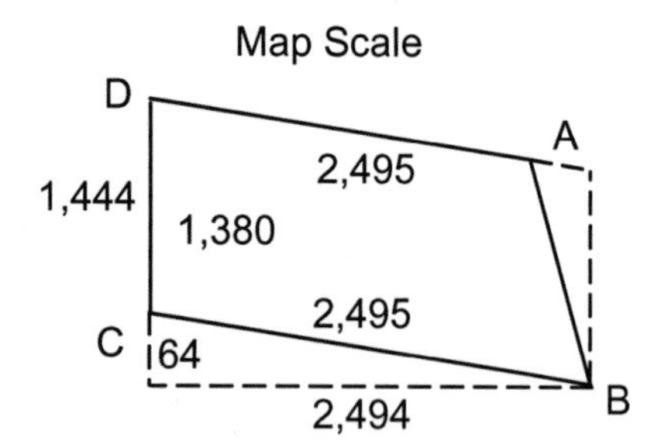

Determine maximum limits of parcel.

Choose scale to place parcel onto a sheet 17 in. × 23 in. (1/2-in. borders).

In the x direction, the largest scale that will fit is to have 2,494 ft represented by 23 in. (1.917 ft). Since 2,494/1.917 = 1,301, this is a scale of 1:1,301.

Similarly, in the y direction, the largest scale that will fit is 1:1,019 [1,444/(17/12)].

The smaller of the two scales (1:1,301) will govern what can be put on the sheet. However, of the choices available, only the scale 1:1,440 will work.

THE CORRECT ANSWER IS: (A)

33. (Reference: See List F)

THE CORRECT ANSWER IS: (B)

34. (Reference: See List A)

1 ft represents 50,000 ft
1 in. represents 4,166 ft
1 in^2 represents 17,361,111 ft^2
10 in^2 represents 173,611,100 ft^2
1 sq. mi. = $(5{,}280 \text{ ft})^2$ = 27,878,400 ft^2

Area = 173,611,100/27,878,400 = 6.23 sq. mi.

THE CORRECT ANSWER IS: (A)

35. (Reference: See List A)

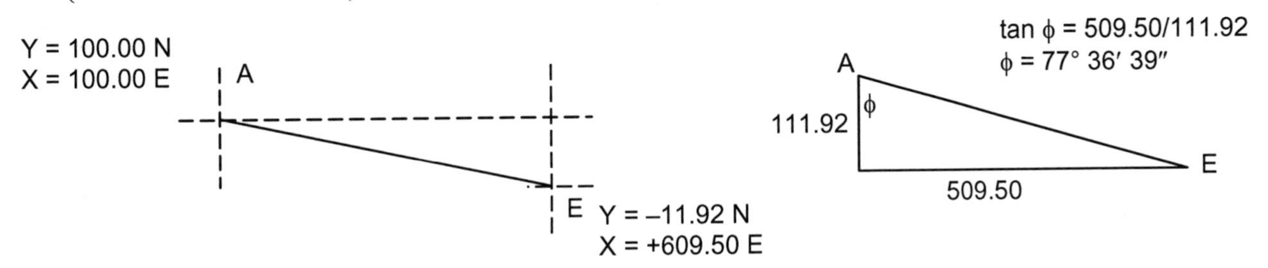

Line AE = 509.50/sin 77°36′39″ (use 77°37′)
AE = 521.56′

THE CORRECT ANSWER IS: (D)

36. (Reference: See List B)

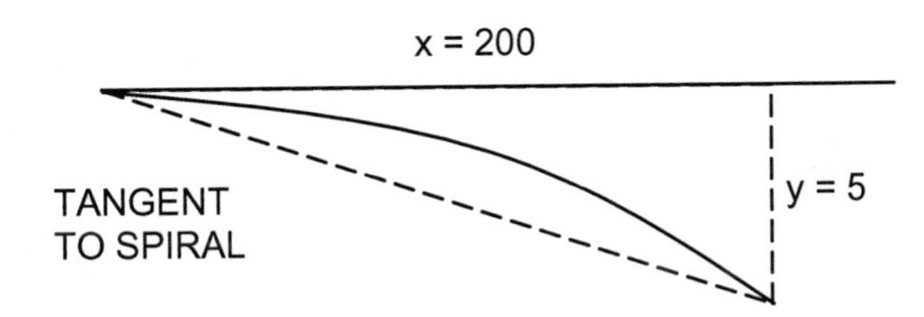

Arctan 5/200 = Deflection angle
= 1°25′55″

THE CORRECT ANSWER IS: (B)

37. (Reference: See List A)

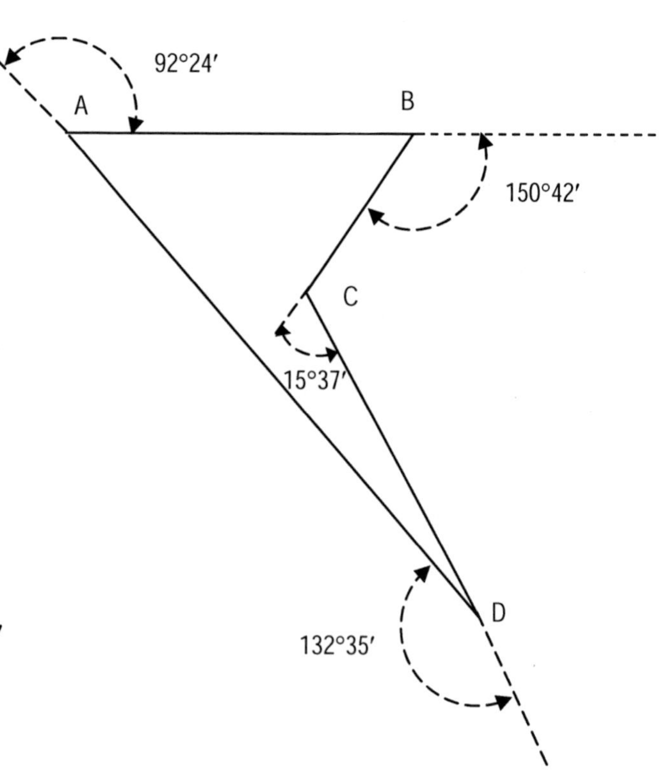

First determine interior angles.
A = 87°36′
B = 29°18′
C = 195°37′
D = 47°25′

Total interior = 359°56′

Error = 04′

Distribute 01′/angle

C = 195°37′ + 01′ = 195°38′

Therefore, balanced angle at C = 15°38′

THE CORRECT ANSWER IS: (C)

38. (Reference: See List A)

D = 36°12′18″
R = 36°12′24″ (corrected)
Mean = 36°12′21″

Therefore D reads 3″ less than actual zenith angle (vertical collimation error).

So: +12°16′12″ elevation angle ⇒ 77°43′48″ actual zenith angle

Subtract 3″ for vertical collimation error.

∴ Set 77°43′45″

THE CORRECT ANSWER IS: (A)

39. (Reference: See List A)

THE CORRECT ANSWER IS: (C)

40. (Reference: See List E)

$$\text{Elevation factor} = \frac{\text{ellipsoidal distance}}{\text{measured distance}}$$

$$= \frac{20{,}906{,}000}{20{,}906{,}000 + \text{elev (ft) above ellipsoid}}$$

$$\text{Sea level distance} = \left(\frac{20{,}906{,}000}{20{,}906{,}000 + 5{,}280}\right)(5{,}280) = 5{,}278.67 \text{ ft}$$

THE CORRECT ANSWER IS: (B)

41. (Reference: See List A)

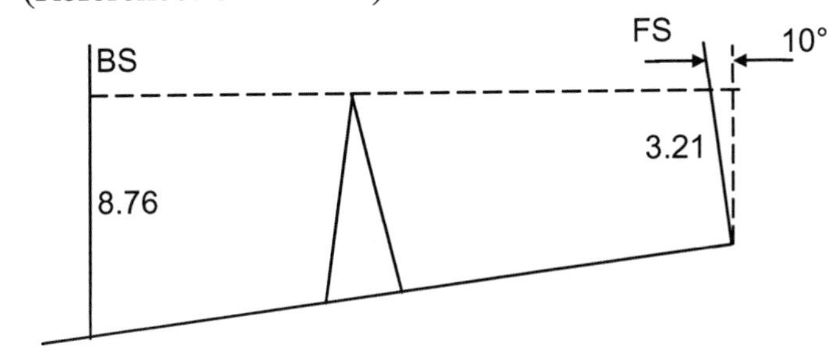

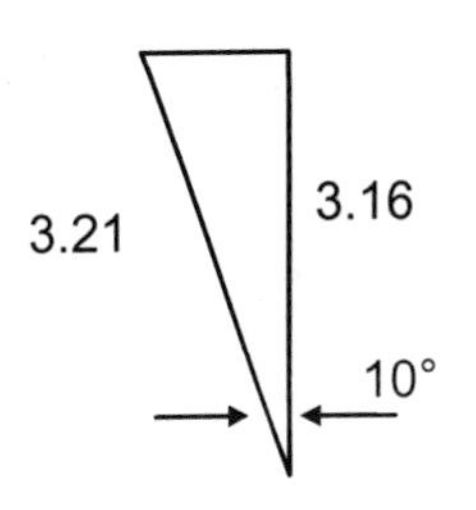

Solve for vertical

V = (3.21)(cos 10°)
V = 3.16
Elev. at B = 644.00 + 8.76 – 3.16 = 649.60 ft

THE CORRECT ANSWER IS: (D)

42. (Reference: See List A)

THE CORRECT ANSWER IS: (A)

43. (Reference: See List F and List H)

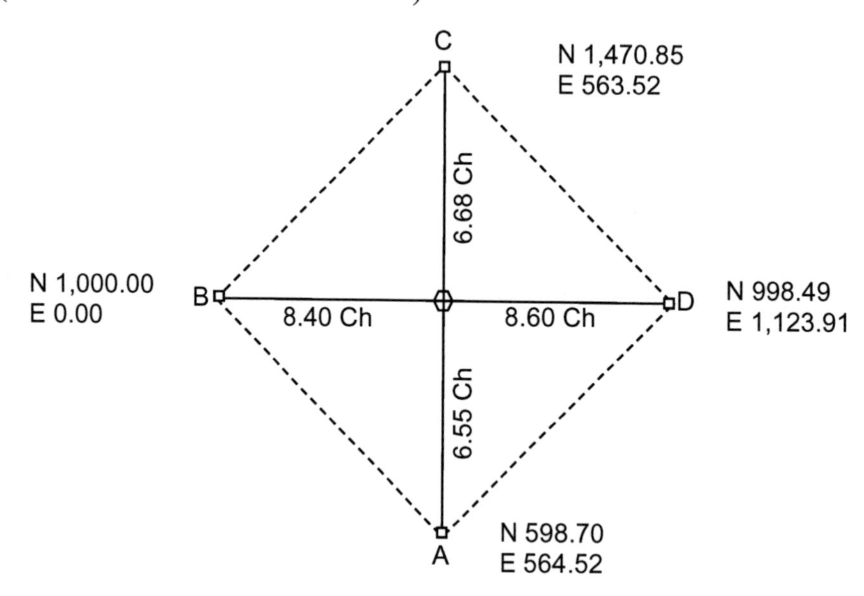

From information in the question, determine coordinates of C, D, and A.

Since B and D (N-S) coordinates are close, the E-W difference = 1,123.91±

Find lost corner E-W coordinate
[8.4/(8.4 + 8.6)](1,123.91) = 555.34

Since A and C (E-W) coordinates are close, the N-S difference = 1,470.85 – 598.70 = 872.15±

Find N-S coordinate:
[6.55/(6.55 + 6.68)](872.15) = 431.79 + 598.70
= 1,030.49

THE CORRECT ANSWER IS: (C)

44. (Reference: See List A)

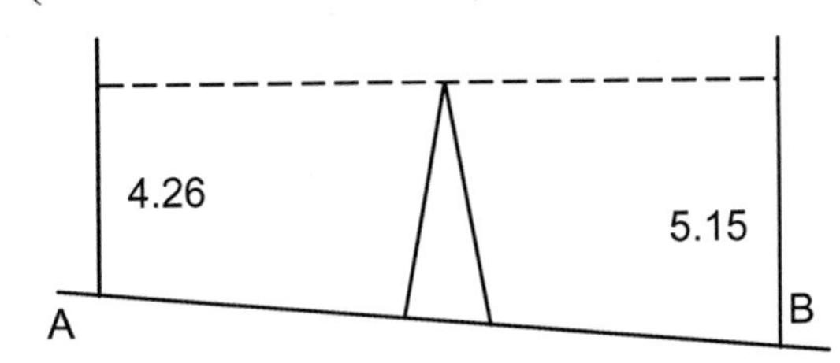

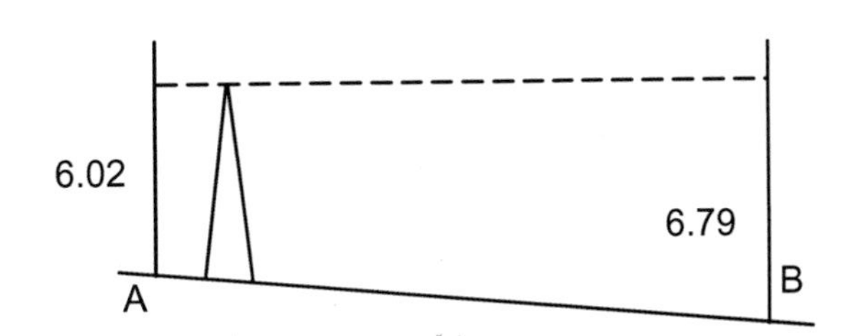

At A: 6.02 – 4.26 = 1.76
At B: 6.79 – 5.15 = 1.64

Difference = 0.12
Adjusted rod reading = 6.79 + 0.12 = 6.91

THE CORRECT ANSWER IS: (D)

45. (Reference: See List F)

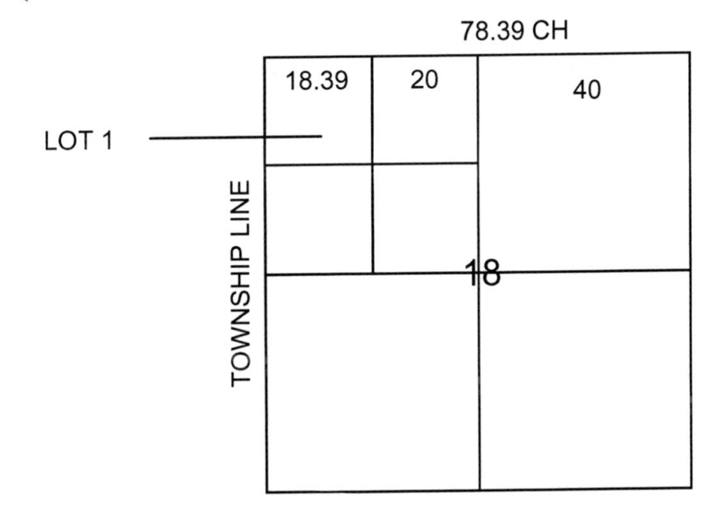

THE CORRECT ANSWER IS: (A)

46. (Reference: See List C)

The basic premise of the "layer model" of GIS design is to separate spatial and attribute data and store them in separate locations and systems.

THE CORRECT ANSWER IS: (A)

47. (Reference: See List B)

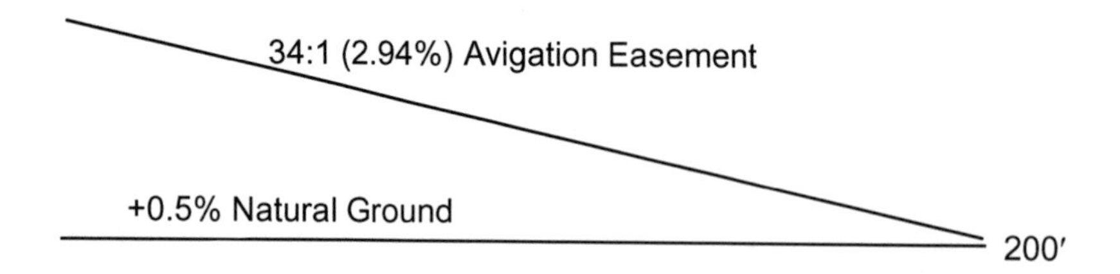

Difference in slope = 2.44%

35/0.0244 = 1,434.4
1,434.4 + 200 ft to runway = 1,634.4 ft

THE CORRECT ANSWER IS: (C)

48. (Reference: See List A)

THE CORRECT ANSWER IS: (B)

49. (Reference: See List A)

THE CORRECT ANSWER IS: (C)

50. (Reference: See List E)

THE CORRECT ANSWER IS: (B)

51. (Reference: See List A and general geometry reference)

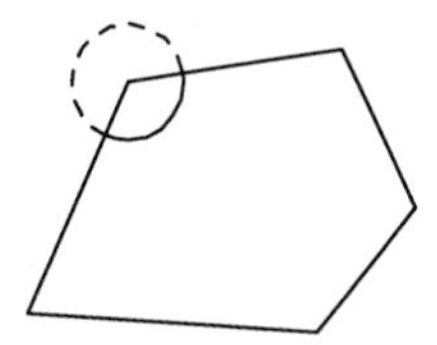

Sum of interior angles = $(n - 2)180$
Total internal and external angles = $360n = 2n(180)$
Sum of exterior angles = $2n(180) - (n - 2)180 = 180(2n - n + 2) = 180(n + 2)$

Sum of exterior = $(n + 2)180 = (8 + 2)180 = 1{,}800°$

THE CORRECT ANSWER IS: (A)

52. (Reference: General mathematics reference)

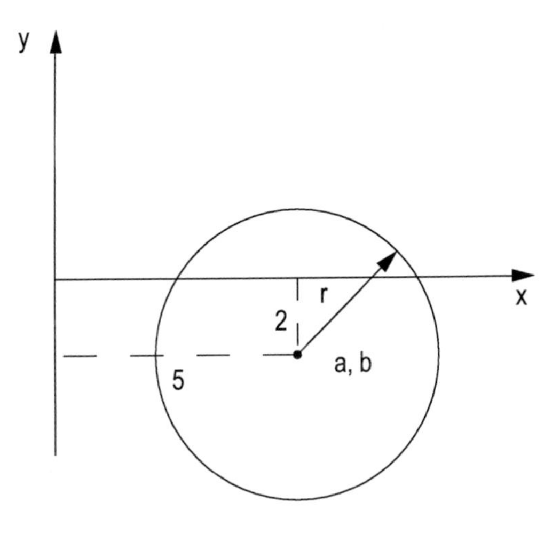

Standard equation of circle:

$(x - a)^2 + (y - b)^2 = r^2$

$\therefore (x - 5)^2 + (y + 2^2) = 16$

THE CORRECT ANSWER IS: (C)

53. (Reference: See List A or a basic statistics reference)

$n = 11$

By calculator, mean = 57.091

Standard deviation, $\sigma = 2.948$

Standard deviation of mean $= \frac{\sigma}{\sqrt{11}} = \frac{2.948}{\sqrt{11}} = 0.889$

THE CORRECT ANSWER IS: (A)

54. (Reference: See List A)

Infrared light is longer wavelength.

THE CORRECT ANSWER IS: (A)

55. (Reference: Basic computing reference)

THE CORRECT ANSWER IS: (C)

56. (Reference: Basic English text)

Atlanta, Georgia, is where the survey was done.

THE CORRECT ANSWER IS: (C)

57. (Reference: Black's Law Dictionary)

THE CORRECT ANSWER IS: (B)

58. (Reference: See List A)

Contents of a total station "log."

THE CORRECT ANSWER IS: (A)

59. (Reference: See List C)

THE CORRECT ANSWER IS: (D)

60. (Reference: See List A)

THE CORRECT ANSWER IS: (D)

61. (Reference: See List A or List B)

$$\begin{array}{r} 157 + 34.55 \\ \underline{-134 + 42.90} \\ 2{,}291.65 \end{array}$$

$2{,}291.65 \times 0.0218 = 49.958$

$$\begin{array}{r} 432.76 \\ \underline{-\,49.958} \\ 382.802 \end{array}$$

THE CORRECT ANSWER IS: (B)

62. (Reference: See List C)

THE CORRECT ANSWER IS: (D)

63. (Reference: See List B)

Without spirals, centripetal force builds up instantaneously at a P.C., causing jerks and damage. Spirals distribute this buildup and decrease the sudden buildup.

THE CORRECT ANSWER IS: (D)

64. (Reference: See List E)

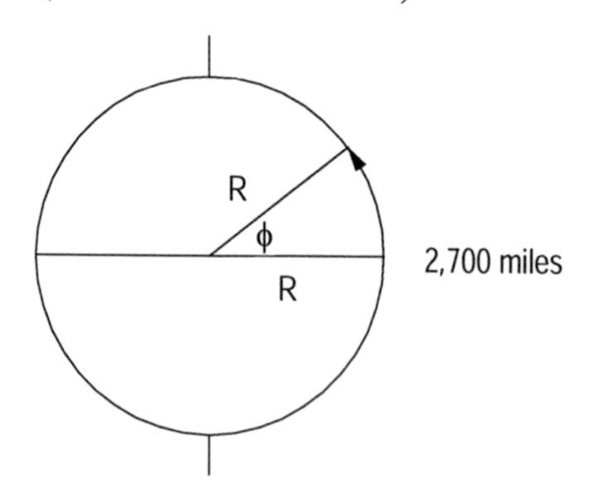

$$\phi = \frac{2,700}{5,631} \times \frac{180}{\pi} = 27.4767°$$
$$\phi = 27°28'21.6''$$

THE CORRECT ANSWER IS: (B)

65. (Reference: General mathematics reference)

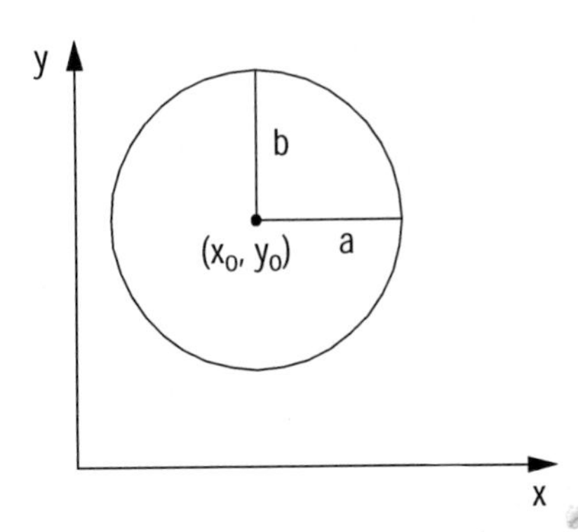

An equation of an ellipse can be cast into the following form:

$$\frac{(x-x_0)^2}{a}+\frac{(y-y_0)^2}{b}=1$$

$$bx^2-2bxx_0+bx_0+ay^2-2ayy_0+ay_0^2=ab$$

x^2 and y^2 terms must be present with positive sign on each

THE CORRECT ANSWER IS: (B)

66. (Reference: See general chemistry or physics text)

Measured temperature	=	46°F
– Correction	=	–3°F
True temperature	=	43°F

$$\text{Celsius} = \frac{5}{9}(43-32)=\frac{5}{9}(11)=6.11°$$

THE CORRECT ANSWER IS: (A)

67. (Reference: See general computer reference)

THE CORRECT ANSWER IS: (D)

68. (Reference: See general reference on grammar and a dictionary)

The spelling mistakes are underlined.

> The deed describes the objects bounding the premices, but parole evidence is usually resorted to, for the purpose of identifying the objects themselves.

Corrections for each mistake are as follows:
1. premises
2. parol

THE CORRECT ANSWER IS: (B)

69. (Reference: See List A or List B)

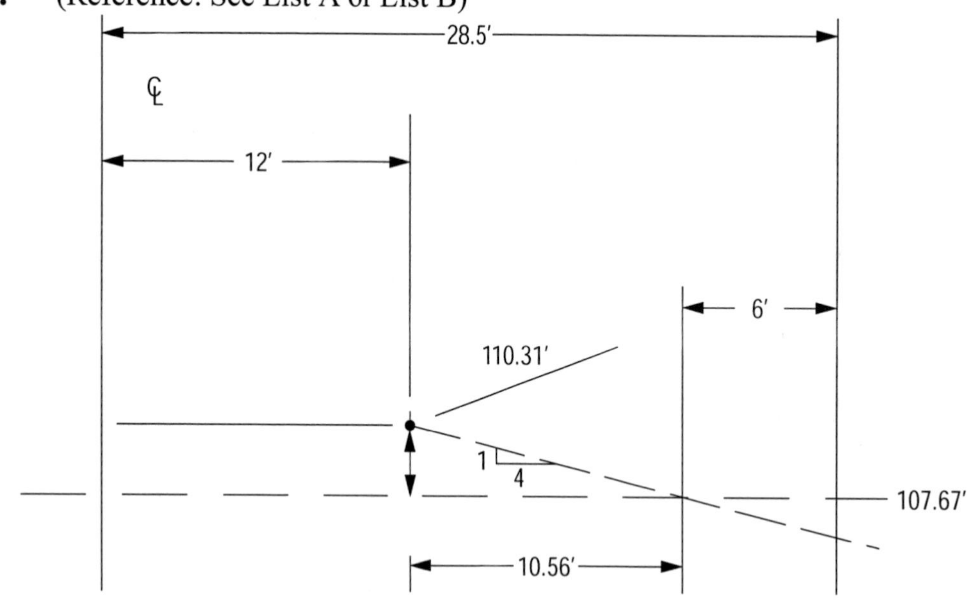

The test point (road edge) is above the ground by $110.31 - 107.67 = 2.64$ ft
The section's in fill $\therefore$ use 4:1 side slope
Move $2.64 \times 4 = 10.56$ right of road edge
From road side, move $28.50 - 22.56 \approx 6$ ft

THE CORRECT ANSWER IS: (A)

70. (Reference: See List C)

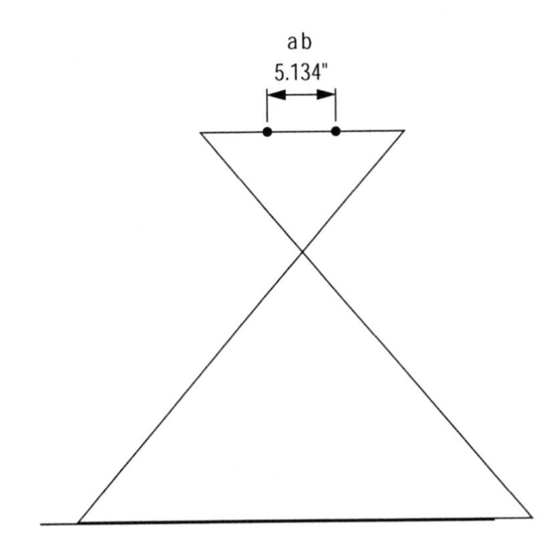

$$AB = \text{ground distance} = \frac{1.689'' \times 24{,}000}{46{,}536 \text{ in.}}$$

$$\text{Scale} = \frac{1}{\left(\frac{AB}{ab}\right)} = \frac{1}{\frac{40.536}{5.134}} = \frac{1}{7{,}896}$$

THE CORRECT ANSWER IS: (B)

71. (Reference: See List A or List B)

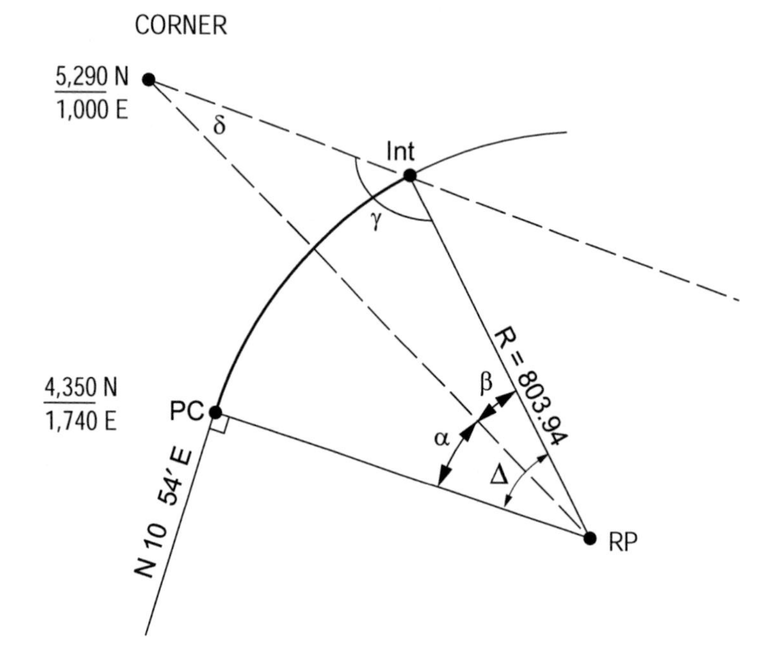

Traverse PC to RP	S 79°06'00" E 803.94	RP = 4,197.98N	2,529.44E
Inverse RP to Corner	Bearing N 54°28'23" W	Distance = 1,879.28	

From bearings $\alpha = 79°06'00'' - 54°28'23''$ — $\alpha = 24°37'37''$

From bearings $\delta = 68°42'00'' - 54°28'23''$ — $\delta = 14°13'36''$

Use Sine Law to solve triangle — RP – Corner – Int

$$\frac{\sin\gamma}{1{,}879.28} = \frac{\sin\delta}{803.94} \qquad \gamma = 144°56'12''$$

and

$$\beta = 180° - (\delta + \gamma) \qquad \beta = 20°50'12''$$

Then for the circular arc

$$\Delta = \alpha + \beta \qquad \Delta = 45°27'49''$$

and

$$L = R\,\Delta\frac{\pi}{180} \qquad L = 637.92$$

THE CORRECT ANSWER IS: (C)

72. (Reference: See List C and List G)

THE CORRECT ANSWER IS: (B)

73. (Reference: See List A and B)

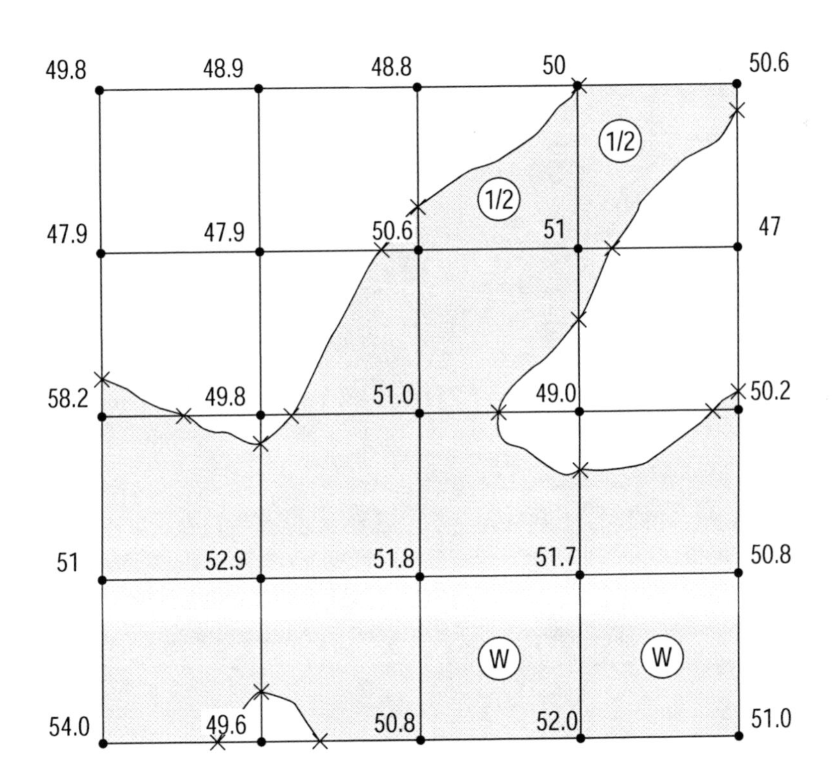

Area above 650 (by inspection)

35% shaded area over half
50% shaded area over half
65% just about right
80% shaded are not 80%

THE CORRECT ANSWER IS: (C)

74. (Reference: See List A, List D, or general trigonometry reference)

$1/2 \text{ in.} = 1/2(1/12) \text{ ft}$

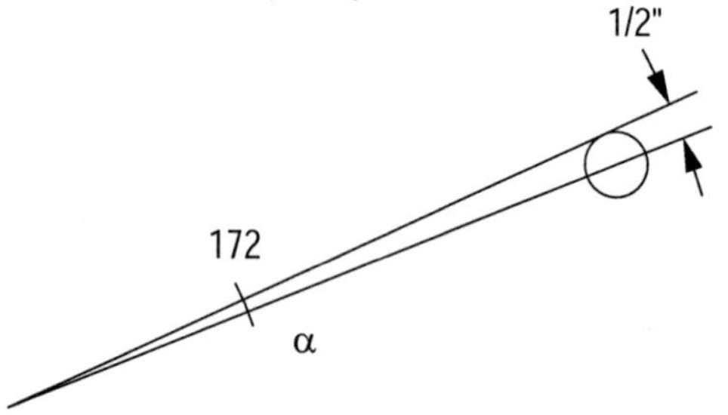

1/2"

172

α

$$\alpha = \tan^{-1} \frac{(1/2)1/12}{172} = 0°00'50''$$

THE CORRECT ANSWER IS: (B)

75. (Reference: General calculus reference)

The cross section area of the tank is a circle given by $x = 10 - \sqrt{y}$.

The area of the circle is $A = \pi x^2$.

Integrating over the height of the tank

$$V = \int_0^{25} A dy = \int_0^{25} \pi x^2 dy = \int_0^{25} \pi\left(10 - \sqrt{y}\right)^2 dy$$

$$= \pi \int_0^{25} \left(100 - 20\sqrt{y} + y\right) dy$$

$$= \pi \left[100y - \frac{2(20)}{3} y^{3/2} + \frac{y^2}{2}\right]_0^{25}$$

$$= \pi \left[100(25) - \frac{2(20)25^{3/2}}{3} + \frac{25^2}{2}\right]$$

$$= \frac{6,875}{6} \pi \cong 3,600$$

THE CORRECT ANSWER IS: (D)

76. (Reference: See List A or general physics reference)

Vertical refraction bends light concave down.

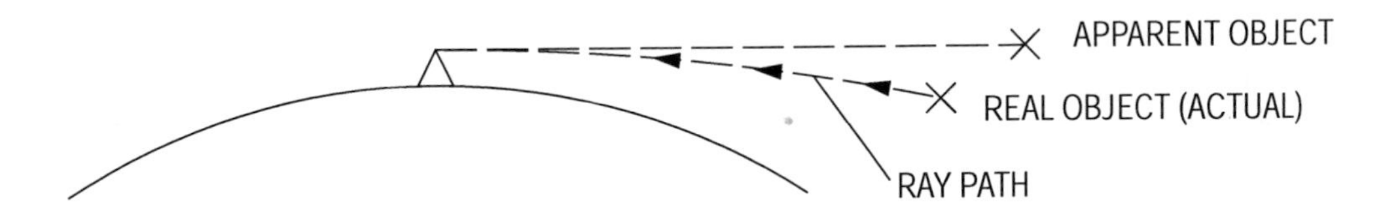

OBJECTS APPEAR HIGHER THAN ACTUAL.

THE CORRECT ANSWER IS: (A)

77. (Reference: General computer reference)

1 byte = 8 bits

THE CORRECT ANSWER IS: (A)

78. (Reference: See List C)

Line on a photo containing nadir point and optical axis of lens.

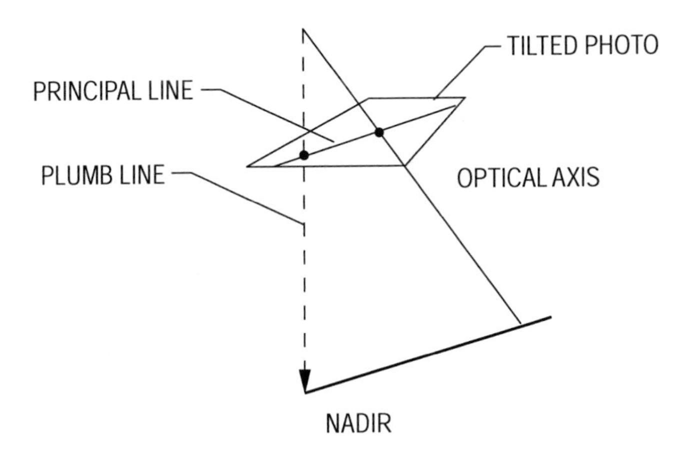

THE CORRECT ANSWER IS: (D)

79. (Reference: See List A or List B)

Highway curve
$D = 40°$
$I(B) = 120°$
$PC = 9+60$

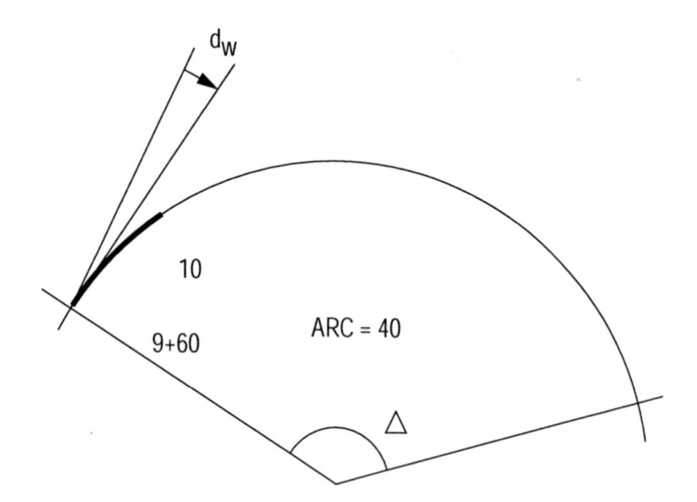

$$d_{100} = \frac{D}{2} = 20°$$
$$d_{40} = 0.40(20°) = 8°$$

THE CORRECT ANSWER IS: (D)

80. (Reference: See List A)

Sta.	10'	20'	21'	26'	30'
1 + 100	792.4	790.1			790.1
+75	792.4	790.0	789.95	789.7	789.5
+50	792.4	789.9			788.9

Elev @ +75 21'

$790.00 - 789.5 = 0.5$

$0.5/10 = 0.05 \quad (0.05)1 = 0.05$

$790.0 - 0.05 = 789.95$

Elev @ +75 26'

$(0.05)6 = 0.30$

$790.00 - 0.30 = 789.70$

$\text{Slope} = \frac{\text{rise}}{\text{run}}$ rise = 789.95 – 789.7 = 0.25

–0.25 (going down)

run = 5'

$$\text{Slope} = \frac{-0.25}{5} = -0.05\%$$

THE CORRECT ANSWER IS: (B)

81. (Reference: See List B)

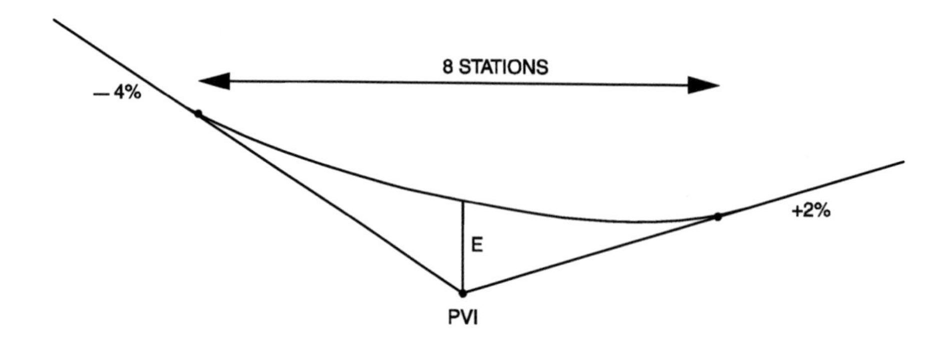

$$r = \frac{g_2 - g_1}{L} = \frac{+2-(-4)}{8} = \frac{3}{4}\ \%/\text{sta}$$

$$E = a\left(\frac{L}{2}\right)^2 = \frac{r}{2}\left(\frac{L}{2}\right)^2 = \frac{3}{4(2)}\left(\frac{8}{2}\right)^2 = 6.0$$

THE CORRECT ANSWER IS: (D)

82. (Reference: General trigonometry reference)

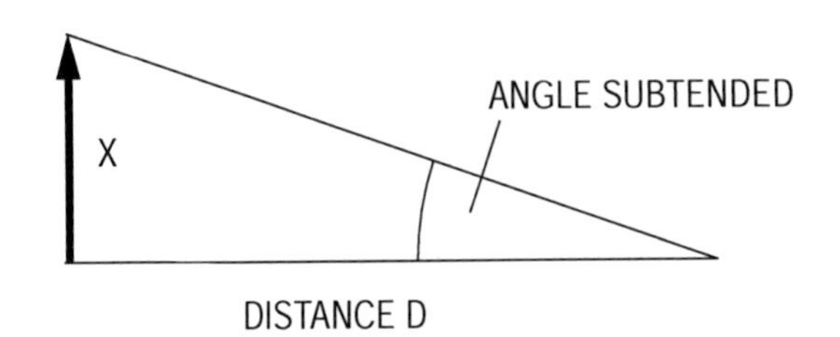

Angle subtended = arctan (X/D)

Object	X – "Width or Height"	D – Distance Away	Angle
Tree	18 ft	100 × 3 = 300 ft	3.433°
House	12	180	3.814°
Coin	0.5 in.	10 in.	2.862°
Moon	2,170 miles	240,000 miles	0.518°

The house subtends the greatest angle = 3.814°.

THE CORRECT ANSWER IS: (B)

83. (Reference: General physics reference)

Red light has a longer wavelength than blue light.

THE CORRECT ANSWER IS: (D)

84. (Reference: General computer reference)

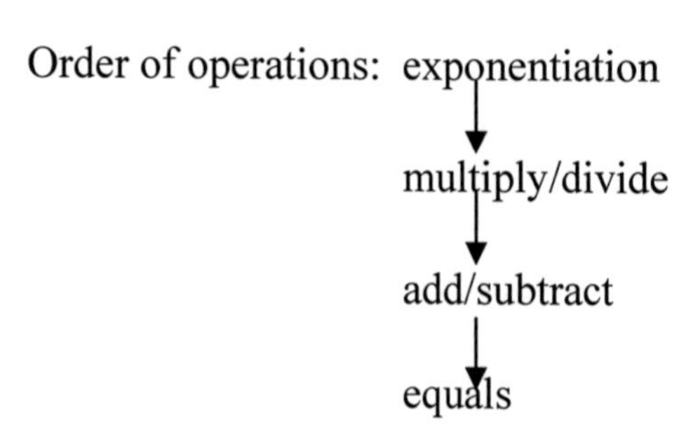

$$A = B * C + D/C \wedge 2$$

$$\left\{ \begin{array}{l} B = 2 \\ C = 0.5 \\ D = 0.27 \end{array} \right\}$$

is same as $\quad A = B * C + \dfrac{D}{C^2}$

$$= 2 \times 0.5 + \frac{127}{0.5^2}$$

$$= 1 + \frac{127}{0.25} = 509$$

THE CORRECT ANSWER IS: (A)

85. (Reference: See List A)

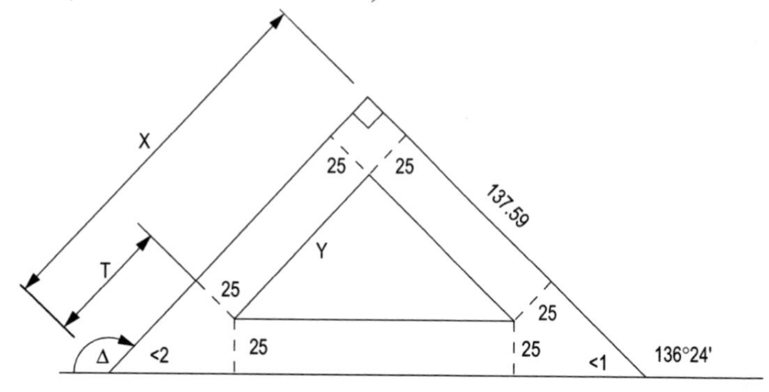

$$\angle 1 = 180° - 136°24' = 43°36'$$

$$\angle 2 = 90 - \angle 1 = 46°24'$$

$$\Delta = 180 - \angle 2 = 133°36'$$

$$T = 25 \tan \frac{133°36'}{2} = 58.329'$$

$$\tan \angle 1 = \frac{x}{137.59}, \qquad x = 131.025'$$

$$y = 131.025 - 25 - 58.329 = 47.696'$$

THE CORRECT ANSWER IS: (C)

SURVEYING STUDY REFERENCES

Other books are suitable references if they contain substantially the same content as those listed. The list is assembled for guide purposes only and is not intended to be exhaustive. Inclusion on the list should not be considered an endorsement.

A. General Surveying Reference Books

Elementary Surveying, An Introduction to Geomatics, 10th edition, 2002, P. R. Wolf and C. Ghilani, Prentice Hall, Upper Saddle River, NJ

Geomatics, 2003, B. Kavanagh, Prentice Hall, Upper Saddle River, NJ

Surveying Theory and Practice, 7th edition, 1998, J. Anderson, and E. Mikhail, McGraw-Hill, New York, NY

Surveying, 9th edition, 1992, Moffitt, Bouchard/10th edition (or later edition), 1997, Harper/Collins, Glenview IL, Moffitt, Bossler

B. Route and Construction Surveying

Surveying with Construction Applications, 4th edition, 2001, B. Kavanagh, Prentice Hall, Upper Saddle River, NJ

Route Surveying and Design, 5th edition (or later edition), 1981, C.F. Meyer, D. W. Gibson, Harper & Row, New York, NY

Surveying: Principles and Applications, 4th edition (or later edition), 1996, Prentice Hall, Englewood Cliffs, NJ, B. Kavanaugh, S.J. Glenn Bird

C. GIS, Mapping, Photogrammetry

Concepts and Techniques of Geographic Information Systems, C.P Lo and A.K.W. Yeung, 2002, Prentice Hall, Upper Saddle River, NJ

Geographic Information Systems: An Introduction, 2002, T. Bernhardsen, John Wiley & Sons, New York, NY

GIS Fundamentals, A First Text on Geographic Information Systems, 2002, P. Bolstad, Eider Press, White Bear Lake, MN

Introduction to Modern Photogrammetry, 2001, Edward M. Mikhail, James S. Bethel, and J. Chris McGlone, John Wiley & Sons, Inc., New York, NY

Elements of Photogrammetry with Applications in GIS, 3rd edition, 2000, Paul Wolf and Bon A. Dewitt, McGraw-Hill, New York, NY

Understand GIS, 3rd edition (or later edition), 1995, by E.S.R.I., Harlow, Essex, England: Longman Scientific and Technical, New York, NY

Geographic Information Systems, A Guide to the Technology, 1st edition (or later edition), 1991, Antenucci, Brown, Croswell and Kevany

An Introduction to Urban Geographic Information Systems, 1st edition (or later edition), 1991 Huxhold, Oxford University Press, New York, NY

D. Measurement Errors and Adjustment Computations

Adjustment Computations, Statistics and Least Squares in Surveying and GIS, 1997, P. R. Wolf and C. Ghilani, John Wiley & Sons, New York, NY

Observations and Least Squares, 1st edition (or later edition) 1976, Mikhail, New York, IEP

E. Geodesy and GPS

State Plane Coordinate System of 1983, 1991, J. Stem, National Geodetic Survey, Rockville, MD

Geodesy, 2nd edition, 1991, W. Torge, Walter de Gruyter and Company, Berlin

Geodesy, any edition, Bomford, Oxford, Clarendon Press

GPS Satellite Land Surveying, 3rd edition, 1995, Leick, John Wiley, Hoboken, NJ

F. Boundary and Cadastral Surveying

Law of Surveying and Boundaries, 5th edition, 1987, W. Robillard and L. Bouman, The Michie Company, Charlottesville, VA

Writing Legal Descriptions in Conjunction with Survey Boundary Control, 1979, G. Wattles, Wattles Publications, Tustin, CA

Evidence and Procedures for Boundary Locations, 2nd edition, 1981, C. Brown, W. Robillard, and D. Wilson, John Wiley & Sons, New York, NY

Brown's Boundary Control and Legal Principles, 4th edition, 1995, C. Brown, W. Robillard, D. Wilson, G. Cole, and F. Uzes, John Wiley & Sons, New York, NY

Land Survey Systems, 1978, J. McEntyre, John Wiley & Sons, New York, NY

Advanced Land Descriptions, 1993, P. Cuomo and R. Minnick, Landmark Enterprises, Rancho Cordova, CA

G. Land Development

Practical Manual of Land Development, 3rd edition, 1998, Colley, McGraw-Hill, New York, NY

Land Development Handbook: Planning, Engineering, and Surveying, 1996, Dewberry and Davis, McGraw-Hill, New York, NY

Land Development for Civil Engineers, 1993, T. Dion, John Wiley & Sons, New York, NY

H. Survey Procedures, Standards, and Dictionaries

Black's Law Dictionary, 2nd edition, 2001, B. Garner, West Group, St. Paul, MN

The Surveying Handbook, 2nd edition (or later edition) 1995, Brinker, Minnick, Chapman & Hall, New York, NY

Definitions of Surveying and Associated Terms, 1978, ACSM and ASCE,

ALTA/ACSM Land Title Survey Standards, 1999 (or later versions)

Land Development Handbook Planning, Engineering and Surveying, 1996, Dewberry & Davis, McGraw-Hill, New York, NY

Manual of Instructions for the Survey of the Public Lands of the United States, 1973, Bureau of Land Management, Technical Bulletin 6, U.S. Dept. of Interior

APPENDIX A

SAMPLE OF EXAM COVERS AND INSTRUCTIONS

Serial No. **SAMPLE**

Name: ______________________________
Last First Middle Initial

Fundamentals of Surveying

MORNING SESSION

READ THE FOLLOWING INSTRUCTIONS CAREFULLY.

- If you do not comply with these instructions, your examination score may be **INVALIDATED.**
- Darken **ALL** answers on the answer sheet enclosed within this examination book. Only the answer sheet will be scored. The answer sheet is the only record of your answers.
- **COMPLETELY DARKEN** the circles corresponding to the answers you choose. Use **ONLY** the pencil provided.
- Do all scratch work only in the blank spaces in this examination book. You may **NOT** write on loose paper. **NO CREDIT** will be given for any work written in the examination book. You will **NOT** be given extra time to transfer answers to the answer sheet.
- This is a closed-book examination. No reference materials may be used. Reference formulas appear on pages 1–4.
- **NOTE:** Unless a question states otherwise, the following apply to this examination:
  - ➢ All azimuths are measured from the north.
  - ➢ X coordinates increase to the east.
  - ➢ Y coordinates increase to the north.
  - ➢ Angular measurements are in degrees (°), minutes ('), and seconds ("), except where decimals may be used as noted.
- Devices or materials that might compromise the security of the examination or examination process are **NOT PERMITTED.** Calculators with communication or text-editing capabilities are prohibited, as are communication devices such as pagers and cellular phones. Calculating and computing devices having a QWERTY keypad arrangement similar to a typewriter are not permitted. These devices include but are not limited to palmtop, laptop, and handheld computers, calculators, databanks, data collectors, and organizers. Prohibited items will be collected by a proctor.
- Copying examination questions for future reference, recording examination questions into an electronic device, copying answers from other examinees, or cheating of any kind is **NOT PERMITTED** and will result in an invalidation of your examination score.
- The morning examination is a maximum of four hours in length. Work all **85** questions according to the instructions. Select the **best** answer from a list of four choices. Each question has equal weight; points are not subtracted for incorrect responses. It is to your advantage to answer every question.
- Only one answer is permitted for each question; no credit is given for multiple answers. If you change an answer, be sure to completely erase the previous mark. Incomplete erasures may be read as intended answers.
- At the conclusion of the examination, you are responsible for returning the numbered examination book that was assigned to you.

SAMPLE

Fundamentals of Surveying—Morning Session

INSTRUCTIONS FOR COMPLETING ANSWER SHEET

1. When the proctor prompts you, slide the answer sheet out from under the front cover of this examination book. **DO NOT** break the seal on the examination book. **DO NOT** open the examination book.

2. Box 1 requests your examination book serial number. It is found on the top front of this examination book.

3. Complete the information requested in box 2. Print legibly and neatly so that the answer sheet can be identified.

4. Box 3 contains a very important agreement between you and NCEES and your local licensure board. If you decide not to sign the agreement, raise your hand and a proctor will collect your examination materials uncompleted. By signing the agreement, you are legally bound to abide by the terms of the agreement. If you do not abide by the terms, your exam score may be invalidated, and/or you may be barred from retaking the exam. These terms include affirmation that the answers you provide are solely of your knowledge and hand; you will not copy any information onto material to be taken from the exam room; you will not reveal in whole or in part any exam questions, answers, problems, or solutions to anyone during or after the exam, whether orally, in writing, or in any Internet "chat rooms" or otherwise. Once you have read, understood, and agreed to the terms of the agreement, sign your name in box 3.

5. Complete the information requested in boxes 4 and 5. This information is very important for identification.

6. Box 6 requests your examinee identification (I.D.) number. Incorrectly entering your I.D. number may delay your score. If necessary, add leading zeros to your I.D. number so that every box is filled. Do not include letters in your I.D. number. For example, if your I.D. number is T1580, enter the number like this:

0	0	0	0	0	1	5	8	0

7. Complete the information requested in box 7. If the day of your birth is one digit, add a zero in front of the digit so that all boxes are filled. This information is very important for identification purposes.

8. You will not receive additional time to transfer answers to the answer sheet. When the proctor says the exam has ended, you must put down your pencil and stop writing.

9. If you have a concern regarding the validity of an examination question, request an Examinee Comment Form from a proctor as you exit the room after you have completed the examination.

DO NOT OPEN THE EXAMINATION BOOK UNTIL INSTRUCTED TO DO SO BY THE PROCTOR.

Serial No. SAMPLE

Name: ____
Last First Middle Initial

Fundamentals of Surveying

AFTERNOON SESSION

READ THE FOLLOWING INSTRUCTIONS CAREFULLY.

- If you do not comply with these instructions, your examination score may be **INVALIDATED.**
- Darken **ALL** answers on the answer sheet enclosed within this examination book. Only the answer sheet will be scored. The answer sheet is the only record of your answers.
- **COMPLETELY DARKEN** the circles corresponding to the answers you choose. Use **ONLY** the pencil provided.
- Do all scratch work only in the blank spaces in this examination book. You may **NOT** write on loose paper. **NO CREDIT** will be given for any work written in the examination book. You will **NOT** be given extra time to transfer answers to the answer sheet.
- This is a closed-book examination. No reference materials may be used. Reference formulas appear on pages 1–4.
- **NOTE:** Unless a question states otherwise, the following apply to this examination:
  - All azimuths are measured from the north.
  - X coordinates increase to the east.
  - Y coordinates increase to the north.
  - Angular measurements are in degrees (°), minutes ('), and seconds ("), except where decimals may be used as noted.
- Devices or materials that might compromise the security of the examination or examination process are **NOT PERMITTED**. Calculators with communication or text-editing capabilities are prohibited, as are communication devices such as pagers and cellular phones. Calculating and computing devices having a QWERTY keypad arrangement similar to a typewriter are not permitted. These devices include but are not limited to palmtop, laptop, and handheld computers, calculators, databanks, data collectors, and organizers. Prohibited items will be collected by a proctor.
- Copying examination questions for future reference, recording examination questions into an electronic device, copying answers from other examinees, or cheating of any kind is **NOT PERMITTED** and will result in an invalidation of your examination score.
- The afternoon examination is a maximum of four hours in length. Work all **85** questions according to the instructions. Select the **best** answer from a list of four choices. Each question has equal weight; points are not subtracted for incorrect responses. It is to your advantage to answer every question.
- Only one answer is permitted for each question; no credit is given for multiple answers. If you change an answer, be sure to completely erase the previous mark. Incomplete erasures may be read as intended answers.
- At the conclusion of the examination, you are responsible for returning the numbered examination book that was assigned to you.

SAMPLE

Fundamentals of Surveying—Afternoon Session

INSTRUCTIONS FOR COMPLETING ANSWER SHEET

1. When the proctor prompts you, slide the answer sheet out from under the front cover of this examination book. **DO NOT** break the seal on the examination book. **DO NOT** open the examination book.

2. Box 1 requests your examination book serial number. It is found on the top front of this examination book.

3. Complete the information requested in box 2. Print legibly and neatly so that the answer sheet can be identified.

4. Box 3 contains a very important agreement between you and NCEES and your local licensure board. If you decide not to sign the agreement, raise your hand and a proctor will collect your examination materials uncompleted. By signing the agreement, you are legally bound to abide by the terms of the agreement. If you do not abide by the terms, your exam score may be invalidated, and/or you may be barred from retaking the exam. These terms include affirmation that the answers you provide are solely of your knowledge and hand; you will not copy any information onto material to be taken from the exam room; you will not reveal in whole or in part any exam questions, answers, problems, or solutions to anyone during or after the exam, whether orally, in writing, or in any Internet "chat rooms" or otherwise. Once you have read, understood, and agreed to the terms of the agreement, sign your name in box 3.

5. Complete the information requested in boxes 4 and 5. This information is very important for identification.

6. Box 6 requests your examinee identification (I.D.) number. Incorrectly entering your I.D. number may delay your score. If necessary, add leading zeros to your I.D. number so that every box is filled. Do not include letters in your I.D. number. For example, if your I.D. number is T1580, enter the number like this:

0	0	0	0	0	1	5	8	0

7. Complete the information requested in box 7. If the day of your birth is one digit, add a zero in front of the digit so that all boxes are filled. This information is very important for identification purposes.

8. Complete the information requested in box 8. This information is very important for statistical purposes and has no effect on scoring.

9. You will not receive additional time to transfer answers to the answer sheet. When the proctor says the exam has ended, you must put down your pencil and stop writing.

10. If you have a concern regarding the validity of an examination question, request an Examinee Comment Form from a proctor as you exit the room after you have completed the examination.

DO NOT OPEN THE EXAMINATION BOOK
UNTIL INSTRUCTED TO DO SO BY THE PROCTOR.

APPENDIX B
SAMPLE ANSWER SHEETS

SAMPLE

NATIONAL COUNCIL OF EXAMINERS FOR ENGINEERING AND SURVEYING

FUNDAMENTALS OF LAND SURVEYING EXAMINATION, AM

Mark Reflex® by NCS EM-161254-2:654321 ED05 Printed in U.S.A.

1 EXAMINATION BOOKLET SERIAL NUMBER

2 Please PRINT your NAME below:

Last name First name Middle initial

TEST DATE: Month Day Year

LOCATION: City State

3 PLEASE READ, THEN SIGN YOUR NAME BELOW:

I affirm by my signature below that I am the person taking this exam, the answers contained hereon are solely of my knowledge and hand, and I have not taken this exam in the previous 30 days.

I further affirm that I will not copy any information onto material to be taken from the exam room. Nor will I reveal in whole or in part any exam questions, answers, problems or solutions to anyone during or after the exam, whether orally, in writing, or any internet "chat rooms," or otherwise. I understand that failure to comply with this statement could result in an invalidation of my exam results and/or bar me from retaking the exam for a time at the discretion of the board.

Signature

4 BOARD CODE

Ala., Alaska, Ariz., Ark., Calif., Colo., Conn., Del., D.C., Fla., Ga., Guam, Hawaii, Idaho, Ill., Ind., Iowa, Kans., Ky., La., Maine, Md., Mass., Mich., Minn., Miss., Mo., Mont., MP, Nebr., Nev., N.H., N.J., N. Mex., N.Y., N.C., N. Dak., Ohio, Okla., Oreg., Pa., P.R., R.I., S.C., S. Dak., Tenn., Tex., Utah, Vt., V.I., Va., Wash., W. Va., Wis., Wyo.

5 LAST NAME First 4 Letters / 1st INIT.

A B C D E F G H I J K L M N O P Q R S T U V W X Y Z

6 EXAMINEE IDENTIFICATION NUMBER

0 1 2 3 4 5 6 7 8 9

7 DATE OF BIRTH

MONTH: Jan, Feb, Mar, Apr, May, Jun, Jul, Aug, Sept, Oct, Nov, Dec

DAY: 0 1 2 3 / 0–9

19-YEAR: 0–9

INSTRUCTIONS

USE NO. 2 PENCIL ONLY

- Do NOT use ink or ballpoint pen
- Erase completely any marks you wish to change
- Make NO stray marks on this answer sheet
- Incomplete erasures and stray marks may be read as intended answers
- Make heavy black marks that completely fill the circle

IMPROPER MARKS PROPER MARK

EXAM

1–85: Ⓐ Ⓑ Ⓒ Ⓓ

SAMPLE

NATIONAL COUNCIL OF EXAMINERS FOR ENGINEERING AND SURVEYING

FUNDAMENTALS OF LAND SURVEYING EXAMINATION, PM

Mark Reflex® by NCS EM-161256-4:654321 ED05 Printed in U.S.A.

1 EXAMINATION BOOKLET SERIAL NUMBER

2 Please PRINT your NAME below:

Last name First name Middle initial

TEST DATE: Month Day Year

LOCATION: City State

3 PLEASE READ, THEN SIGN YOUR NAME BELOW:

I affirm by my signature below that I am the person taking this exam, the answers contained hereon are solely of my knowledge and hand, and I have not taken this exam in the previous 30 days.

I further affirm that I will not copy any information onto material to be taken from the exam room. Nor will I reveal in whole or in part any exam questions, answers, problems or solutions to anyone during or after the exam, whether orally, in writing, or any internet "chat rooms," or otherwise. I understand that failure to comply with this statement could result in an invalidation of my exam results and/or bar me from retaking the exam for a time at the discretion of the board.

Signature

BIOGRAPHICAL DATA FROM BOX #8 (BELOW) IS GATHERED FOR STATISTICAL PURPOSES AND HAS NO EFFECT ON SCORING.

4 BOARD CODE

○ Ala. ○ Alaska ○ Ariz. ○ Ark. ○ Calif. ○ Colo. ○ Conn. ○ Del. ○ D.C. ○ Fla. ○ Ga. ○ Guam ○ Hawaii ○ Idaho ○ Ill. ○ Ind. ○ Iowa ○ Kans. ○ Ky. ○ La. ○ Maine ○ Md. ○ Mass. ○ Mich. ○ Minn. ○ Miss. ○ Mo. ○ Mont. ○ MP ○ Nebr. ○ Nev. ○ N.H. ○ N.J. ○ N. Mex. ○ N.Y. ○ N.C. ○ N. Dak. ○ Ohio ○ Okla. ○ Oreg. ○ Pa. ○ P.R. ○ R.I. ○ S.C. ○ S. Dak. ○ Tenn. ○ Tex. ○ Utah ○ Vt. ○ V.I. ○ Va. ○ Wash. ○ W. Va. ○ Wis. ○ Wyo.

5 LAST NAME First 4 Letters | 1st INIT.

(A) (B) (C) (D) (E) (F) (G) (H) (I) (J) (K) (L) (M) (N) (O) (P) (Q) (R) (S) (T) (U) (V) (W) (X) (Y) (Z)

6 EXAMINEE IDENTIFICATION NUMBER

(0) (1) (2) (3) (4) (5) (6) (7) (8) (9)

7 DATE OF BIRTH

MONTH | DAY | 19-YEAR

○ Jan ○ Feb ○ Mar ○ Apr ○ May ○ Jun ○ Jul ○ Aug ○ Sept ○ Oct ○ Nov ○ Dec

8

1. How many times have you taken this exam before?
   ○ This is the first time ○ Twice before
   ○ Once before ○ Three or more times before
2. Are you a graduate of or enrolled in a 4 yr. degree program in Surveying, Surveying and Mapping, **Surveying Engineering, or Geomatics?**
   ○ Yes ○ No
3. If YES, from which institution?
   ○ California State Polytechnic Univ., Pomona
   ○ California State Univ., Fresno
   ○ East Tennessee State Univ.
   ○ Ferris State Univ. - Michigan
   ○ Michigan Technological Univ.
   ○ New Jersey Institute of Tech.
   ○ New Mexico State Univ.
   ○ Ohio State Univ.
   ○ Oregon Institute of Tech.
   ○ Purdue Univ. - Indiana
   ○ State Univ. of New York - Alfred
   ○ Univ. of Alaska - Anchorage
   ○ Univ. of Florida
   ○ Univ. of Maine
   ○ Univ. of Wisconsin - Madison
   ○ Other than listed
4. If you answered NO to 2, do any of the following apply?
   ○ 4 yr. degree in Engineering
   ○ Other 4 yr. degree
   ○ 2 yr. degree
   ○ No degree

INSTRUCTIONS

USE NO. 2 PENCIL ONLY

- Do NOT use ink or ballpoint pen
- Erase completely any marks you wish to change
- Make NO stray marks on this answer sheet
- Incomplete erasures and stray marks may be read as intended answers.
- Make heavy black marks that completely fill the circle

IMPROPER MARKS ✓ ✗ ⊙ ◒ PROPER MARK ●

EXAM

86 (A)(B)(C)(D)	103 (A)(B)(C)(D)	120 (A)(B)(C)(D)	137 (A)(B)(C)(D)	154 (A)(B)(C)(D)
87 (A)(B)(C)(D)	104 (A)(B)(C)(D)	121 (A)(B)(C)(D)	138 (A)(B)(C)(D)	155 (A)(B)(C)(D)
88 (A)(B)(C)(D)	105 (A)(B)(C)(D)	122 (A)(B)(C)(D)	139 (A)(B)(C)(D)	156 (A)(B)(C)(D)
89 (A)(B)(C)(D)	106 (A)(B)(C)(D)	123 (A)(B)(C)(D)	140 (A)(B)(C)(D)	157 (A)(B)(C)(D)
90 (A)(B)(C)(D)	107 (A)(B)(C)(D)	124 (A)(B)(C)(D)	141 (A)(B)(C)(D)	158 (A)(B)(C)(D)
91 (A)(B)(C)(D)	108 (A)(B)(C)(D)	125 (A)(B)(C)(D)	142 (A)(B)(C)(D)	159 (A)(B)(C)(D)
92 (A)(B)(C)(D)	109 (A)(B)(C)(D)	126 (A)(B)(C)(D)	143 (A)(B)(C)(D)	160 (A)(B)(C)(D)
93 (A)(B)(C)(D)	110 (A)(B)(C)(D)	127 (A)(B)(C)(D)	144 (A)(B)(C)(D)	161 (A)(B)(C)(D)
94 (A)(B)(C)(D)	111 (A)(B)(C)(D)	128 (A)(B)(C)(D)	145 (A)(B)(C)(D)	162 (A)(B)(C)(D)
95 (A)(B)(C)(D)	112 (A)(B)(C)(D)	129 (A)(B)(C)(D)	146 (A)(B)(C)(D)	163 (A)(B)(C)(D)
96 (A)(B)(C)(D)	113 (A)(B)(C)(D)	130 (A)(B)(C)(D)	147 (A)(B)(C)(D)	164 (A)(B)(C)(D)
97 (A)(B)(C)(D)	114 (A)(B)(C)(D)	131 (A)(B)(C)(D)	148 (A)(B)(C)(D)	165 (A)(B)(C)(D)
98 (A)(B)(C)(D)	115 (A)(B)(C)(D)	132 (A)(B)(C)(D)	149 (A)(B)(C)(D)	166 (A)(B)(C)(D)
99 (A)(B)(C)(D)	116 (A)(B)(C)(D)	133 (A)(B)(C)(D)	150 (A)(B)(C)(D)	167 (A)(B)(C)(D)
100 (A)(B)(C)(D)	117 (A)(B)(C)(D)	134 (A)(B)(C)(D)	151 (A)(B)(C)(D)	168 (A)(B)(C)(D)
101 (A)(B)(C)(D)	118 (A)(B)(C)(D)	135 (A)(B)(C)(D)	152 (A)(B)(C)(D)	169 (A)(B)(C)(D)
102 (A)(B)(C)(D)	119 (A)(B)(C)(D)	136 (A)(B)(C)(D)	153 (A)(B)(C)(D)	170 (A)(B)(C)(D)